ABOUT THE AUTHOR

Jeremy Stranks is a leading authority on health and safety at work, having written and lectured extensively on this subject. His other books for Blackhall Publishing include *Health and Safety at Work in Ireland* and the forthcoming *Food Safety Management in the UK*.

OTHER TITLES IN THE BLACKHALL GUIDES SERIES

Health and Safety at Work in Ireland
Food Safety Management in Ireland
Food Safety Management in the UK
Employment Law in Ireland
Employment Law in the UK

THE BLACKHALL GUIDE TO HEALTH AND SAFETY AT WORK IN THE UK

Jeremy Stranks

BP

BLACKHALL
Publishing

This book was typeset by
Gough Typesetting Services for
BLACKHALL PUBISHING,
26 Eustace Street,
Dublin 2.
(e-mail: blackhall@tinet.ie)

A catalogue record for this book
is available from the British Library

ISBN 1-901657-40-X

The masculine pronoun has been used throughout this book. This
stems from a desire to avoid ugly or cumbersome language, and no discrimination, prejudice or bias is intended.

Printed by
Betaprint Ltd.

PREFACE

Health and safety at work is, for many managers, a difficult subject due to the vast array of legislation that has been brought in over the last decade in particular.

Current health and safety legislation identifies the need for employers to actually manage their health and safety procedures and systems, sooner than merely complying with basic legal requirements.

This book is particularly directed at those managers who may have direct or indirect responsibility for health and safety in their organisations, who may not have the assistance of a health and safety adviser and who need on-the-spot information on a range of issues. 1 hope all those who use this book will find it helpful.

Jeremy Stranks
August 1998

*To Val, Fiona, Simon
and Marguerita*

CONTENTS

PART 4: SAFETY TECHNOLOGY

LIST OF FIGURES

LIST OF TABLES

PART 1

HEALTH AND SAFETY MANAGEMENT

CHAPTER 1

PRINCIPAL LEGAL REQUIREMENTS

CRIMINAL AND CIVIL LIABILITY

Occupational health and safety law is founded on both statute law and the common law.

Breach of a statute (an Act of Parliament) and regulations made under a statute, generally gives rise to criminal liability. This means that an offender can be brought before the criminal courts, e.g. the Magistrates Court and, if found guilty of the offence with which he has been charged, fined or imprisoned, or both. The principal health and safety legislation is the Health and Safety at Work etc. Act 1974 (HSWA) and the Management of Health and Safety at Work Regulations 1992.

A person may also be in breach of a common law duty. The common law is the unwritten law and is based on the decisions of the courts that have been bound by the doctrine of precedent into a body of authoritative principles and rules. Common law is, fundamentally, 'judge-made' law and forms the basis for the law of tort. The torts of negligence and breach of statutory duty have played an important part in the development of civil liability with regard to occupational health and safety. Common law is synonymous with case law and its rules, principles and doctrines are to be found in the law Reports (such as the All England Reports) published on a regular basis.

COMMON LAW

The principal contribution that common law has made to health and safety at work involves the rights of employees, their dependents and other persons to sue an employer for damages for death, personal injury or disease. An employer is expected to take 'reasonable care' for the safety of his employees and other persons affected by his operations. Thus if an employer knows, or ought to have known, of a risk to the health and safety of his employees, he may be liable if an employee dies, is injured or suffers disease as a result of exposure to that risk, or if the employer failed to take reasonable care.

The common law duties of an employer were laid down in *Wilsons & Clyde Coal Co. Ltd v English* [1938] A.C. 57. The common law requires that all employers provide and maintain:

(a) a safe place of work with safe means of access to and egress from same;

(b) safe appliances and equipment and plant for doing the work;

(c) a safe system of work; and

(d) competent and safety-conscious employees.

It is important to recognise that common law precedents frequently form the basis of statute law. In this case, the above common law duties were incorporated in section 2 of the Health and Safety at Work etc. Act 1974 (General duties of employers to their employees)

The above duties apply even though an employee may be working on third party premises, or where an employee has been hired out to another employer, but where the control of the work he is undertaking remains with the permanent employer. The test of whether an employee has been 'temporarily employed' by a second employer is principally one of 'control'.

The Principal Torts

The common law duties of employers listed above form part of the general law of negligence and are linked with the duty of an employer to take reasonable care. A tort is 'a civil wrong' and the principal torts involving health and safety-related issues are negligence and breach of statutory duty.

Negligence

'Negligence' can be broadly defined as "careless conduct injuring another". More specifically, negligence incorporates three elements, namely:

(a) the existence of a duty of care owed by the defendant to the plaintiff;

(b) breach of that duty;

(c) injury, damage or loss resulting from, or caused by, that breach.

(*Lochgelly Iron & Coal Co.Ltd v. M'Mullan* [1934] A.C. 1)

These three elements of negligence must be proved before a person is entitled to bring a claim for damages though, in the case of breach of statutory duty (see below), the plaintiff merely has to show that the breach of that duty was the material cause of his injury.

Breach of Statutory Duty

Breach of a duty under a statute and/or regulations generally gives rise to criminal liability. However, in certain cases, such a breach may give rise to civil liability within the tort of breach of statutory duty. It should be noted that a breach of the general duties under the HSWA (sections 2 to 6) or the

Management of Health and Safety at Work Regulations 1992, does not give rise to civil liability because such liability is specifically excluded in the text of the legislation. However, where such an exclusion is absent from the text of regulations, by implication, civil liability applies.

Thus, a breach of the Workplace (Health, Safety and Welfare) Regulations 1992 gives rise to civil liability. This means, for example, that an employer could, firstly, be prosecuted for breach of these regulations (such as failing to maintain his workplace in efficient state, in efficient working order and in good repair) and, subsequently, be sued by an injured employee whose injuries were directly attributable to that employer's failure to maintain the workplace.

In interpreting Regulations, a check should always be made as to whether a breach of the Regulations in question gives rise to civil liability or otherwise.

The 'Double-barrelled' Action

Because an employee is entitled to sue his employers for damages for injury arising from breach of a common law duty and a statutory duty, this has led to the development of the 'double-barrelled' action. In this type of action, an injured employee sues separately, but simultaneously, for a breach of both duties. (*Kilgollan v. Cooke & Co. Ltd* [1956] 2 All E.R. 294)

STATUTE LAW

The Hierarchy of Duties

Duties under statutes, and regulations made under a statute, may be of an absolute (strict) nature or qualified by the terms 'so far as is practicable' or 'so far as is reasonably practicable'. These terms are significant in the interpretation of the legislation.

Absolute Duties

Where the risk of injury or disease is inevitable if safety requirements are not followed, a statutory duty may well be absolute. An example of an absolute duty on employers arises in Regulation 5(1) of the Provision and Use of Work Equipment Regulations 1992, which states:

> Every employer *shall* ensure that work equipment is so constructed or adapted as to be suitable for the purpose for which it is to be used or provided.

Absolute duties are qualified by the term 'shall' or 'must'.

'Practicable' Requirements

Where a duty is qualified by the term 'so far as is practicable' this implies that if, in the light of current knowledge or invention or in the light of the current state of the art, it is possible to comply with the duty, then, irrespective of the cost or sacrifice involved, such a duty must be complied with. (See *Schwalb v. Fass, H & Son* [1946] 175 I.T. 345.)

'Practicable' means more than 'physically possible' and implies a higher duty of care than a duty qualified by the term 'so far as is reasonably practicable'.

'Reasonably Practicable' Requirements

A duty at this level implies that a computation must be made in which the quantum of risk is placed on one side of the scale and the cost or sacrifice in undertaking the measures necessary for averting the risk is placed on the other side. If it can be shown that there is a gross disproportion between these two factors, that is, the risk is insignificant in relation to the cost or sacrifice, then a defendant discharges the onus on himself. (See *Edwards v. National Coal Board* [1949] 1 A.E.R. 743.) All the duties under HSWA, and in many Regulations, are set at this level.

The Mythical 'Reasonable Man'

The duty of all persons to take 'reasonable care' is well established, but what is a 'reasonable' person? What is it about his behaviour that makes him reasonable and how do the courts interpret this term?

The mythical 'Reasonable Man' was interpreted by one judge in the past as "the man who travels to work every day on the top deck of the No.57 Clapham omnibus". As such, the term is flexible and changes with time, according to society and the norms prevalent at the time.

Under section 7 of HSWA every employee has an absolute duty "while at work to take reasonable care for the health and safety of himself and of other persons who may be affected by his acts or omissions at work".

STATUTES, REGULATIONS, APPROVED CODES AND GUIDANCE NOTES

These are the principal forms of legislation and guidance.

Statutes

These are Acts of Parliament, such as the Factories Act 1961. The principal statute is the Health and Safety at Work etc. Act 1974 (HSWA).

Regulations

The HSWA, as with other statutes, gives the Minister or Secretary of State power to make Regulations (subordinate or delegated legislation). The Regulations may be drafted by the Health and Safety Executive (HSE) and submitted through the Health and Safety Commission (HSC) to the Secretary of State for Employment, such as the Construction (Design and Management) (CDM) Regulations 1994, the Noise at Work Regulations 1989 and the Manual Handling Operations Regulations 1992. Most Regulations arise as a result of the UK implementing European Directives.

There is a general requirement for the HSC and HSE to keep interested parties informed of, and adequately advised on, such matters.

Approved Codes of Practice (ACOPs)

The need to provide further detail and direction on the implementation of Regulations is recognised in section 16 of HSWA which gives the HSC power to prepare and approve *Codes of Practice (ACOP)* on matters contained not only in Regulations, but in sections 2 to 7 of the Act. Before preparing a code, the HSE, acting on behalf of the HSC, must consult with any interested body.

An ACOP is a quasi-legal document and – although non-compliance does not constitute a breach – if the contravention of the Act or Regulations is alleged, the fact that the ACOP was not followed could be accepted in court as evidence of failure to comply with the requirement or to do all that was reasonably practicable, (depending on the level of duty imposed). A defence would be to prove that works of an equivalent nature has been carried out or something equally as good, or better, had been done.

Examples of ACOPs are 'Workplace Health, Safety and Welfare' issued with the Workplace (Health, Safety and Welfare) Regulations 1992 and 'Control of Substances Hazardous to Health' issued with the Control of Substances Hazardous to Health (COSHH) Regulations 1994.

HSE Guidance Notes

The HSE issues *Guidance Notes* to supplement the information in Regulations. Guidance notes have no legal status and are purely of an advisory nature.

There are six categories of Guidance Notes:

- General Safety (GS).

- Chemical Safety (CS).

- Environmental Hygiene (EH).

- Medical Series (MS).

- Plant and Machinery (PM).

- Health and Safety (General) (HS[G]).

Examples of HSE Guidance Notes are:

- EH40   Occupational Exposure Limits.

- MS20   Pre-employment Health Screening.

- PM21   Safety in the Use of Woodworking Machines.

- CS16   Chlorine Vaporisers.

- GS5   Entry into Confined Spaces.

The HSE also issue a wide range of HSE Information Sheets covering areas of such as engineering, construction, wood-working and agriculture.

HEALTH AND SAFETY AT WORK ETC. ACT (HSWA) 1974

The HSWA covers all persons at work, namely employers, the self-employed, employees, controllers of premises and manufacturers, designers of articles and substances etc. used at work, apart from domestic workers in private employment. The Act further extends to the prevention of risks to the health and safety of members of the public.

Objectives of the HSWA

1. To secure the health, safety and welfare of all persons at work.

2. To protect others from the risks arising from workplace activities.

3. To control the obtaining, keeping and use of explosive or highly flammable substances.

4. To control emissions into the atmosphere of noxious or offensive substances.

Most duties are qualified by 'so far as is reasonably practicable'.

General Duties of Employers to their Employees (section 2)

There is a general duty on every employer to ensure, so far as is reasonably practicable, the health, safety and welfare at work of all employees. The matters to which that duty extends include in particular:

(a) the provision and maintenance of plant and systems of work that are, so far as is reasonably practicable, safe and without risks to health;

(b) arrangements for ensuring, so far as is reasonably practicable, safety and absence of risks to health in connection with the use, handling, storage and transport of articles and substances;

(c) the provision of such information, instruction, training and supervision as is necessary to ensure, so far as is reasonably practicable, the health and safety at work of his employees;

(d) so far as is reasonably practicable as regards any place of work under the employer's control, the maintenance of the workplace in a condition that is safe and without risks to health and the provision and maintenance of means of access to and egress from it that are safe and without such risks; and

(e) the provision and maintenance of a working environment for his employees that is, so far as is reasonably practicable, safe, without risks to health and adequate as regards facilities and arrangements for their welfare at work.

Every employer must prepare and, as often as may be appropriate, revise, a written statement of his general policy with respect to the health and safety at work of his employees and the organisation and arrangements for the time being in force for carrying out that policy. He must also bring the statement, and any revision of it, to the notice of all his employees.

Employers have a duty to consult with safety representatives with a view to promoting and developing measures to ensure the health and safety at work of employees and in checking the effectiveness of such measures.

No employer shall levy, or permit to be levied, on any employee of his any charge in respect of anything done or provided in pursuance of any specific requirement of the relevant statutory provisions (section 9).

Duties of Employees (section 7)

It shall be the duty of every employee while at work:

(a) to take reasonable care for the health and safety of himself and of other persons who may be affected by his acts or omissions at work; and

(b) as regards any duty or requirement imposed on his employer or any other person by or under the relevant statutory provisions, to co-operate with him so far as is necessary to enable that duty or requirement to be performed or complied with.

No person shall intentionally or recklessly interfere with or misuse anything provided in the interests of health, safety or welfare in pursuance of any of the relevant statutory provisions (section 8).

Duties of Employers and the Self-employed to Persons other than their Employees (section 3)

Every employer must conduct his undertaking in such a way as to ensure, so far as is reasonably practicable, that persons not in his employment (e.g. contractors' employees), are not exposed to risks to their health or safety.

Similar duties with regard to non-employees are imposed on self-employed persons. Employers and self-employed persons must provide non-employees, who may be affected by the way such persons conduct their undertakings, with information on aspects of the way in which they conduct their undertakings as might affect their health and safety.

Duties of Persons concerned with the Premises to Persons other than their Employees (section 4)

Section 4 has effect for imposing on people, duties in relation to those who:

(a) are not their employees; but

(b) use non-domestic premises made available to them as a place of work or as a place where they may use plant or substances provided for their use there.

It also applies to premises that have been made available and other non-domestic premises used in connection with them.

In this case the person or persons in control of premises must take such measures to ensure, so far as is reasonably practicable, that the premises, all means of access thereto or egress therefrom and any plant or substance in the premises or (as the case may be) provided for use there, is or are safe and without risks to health.

The protection under this section extends to visitors who are:

(a) workers, e.g. employees of a building contractor; and

(b) visitors to premises, e.g. people invited to view historical buildings.

General Duties of Manufacturers etc. as regards Articles and Substances for use at Work (section 6)

Section 6 is concerned with product liability, that area of law concerned with the duties owed by manufacturers, designers of products and other persons in the supply chain, towards all those who use such products. This section originally placed specific duties on designers, manufacturers, importers and suppliers of articles and substances used at work. These duties were amended by the Consumer Protection Act 1987 (CPA) whereby, in the case of articles for use at work, such persons must:

(a) ensure, so far as is reasonably practicable, that the article is so designed and constructed that it will be safe and without risks to health at all times when it is being set, cleaned, used or maintained by a person at work;

(b) carry out or arrange for the carrying out of such testing and examination as may be necessary for the performance of the duty imposed on him by the preceding paragraph;

(c) take such steps as are necessary to secure that persons supplied by that person with the article are provided with adequate information about the use for which the article is designed or has been tested and about any conditions necessary to ensure that it will be safe and without risks to health at all such times as are mentioned in paragraph (a) above and when it is being dismantled or disposed of;

(d) take such steps as are necessary to ensure, so far as is reasonably practicable, that persons so supplied are provided with all such revisions of information provided by them by virtue of the preceding paragraph as are necessary by reason of its becoming known that anything gives rise to a serious risk to health or safety.

In the case of substances for use at work, such persons have a duty to:

(a) ensure, so far as is reasonably practicable, that the substance will be safe and without risks to health at all times when it is being used, handled, processed, stored or transported by a person at work or in premises to which section 4 applies;

(b) carry out or arrange for the carrying out of such testing and examination as may be necessary for the performance of the duty imposed on him by the preceding paragraph;

(c) take such steps as are necessary to secure that persons supplied by that person with the substance are provided with adequate information about any risks to health or safety to which the inherent properties of the substance may give rise, about the results of any relevant tests which have been carried out on or in connection with the substance and about any conditions necessary to ensure that the substance will be safe and without risks to health at all such times as are mentioned in paragraph (a) above and when the substance is being disposed of;

(d) take such steps as are necessary to secure, so far as is reasonably practicable, that persons so supplied are provided with all such revisions of information provided by them by virtue of the preceding paragraph as are necessary by reason of it becoming known that anything gives rise to a serious risk to health or safety.

Under the CPA, 'article for use at work' means:

(a) any plant designed for use or operation (whether exclusively or not) by persons at work or who erect or install any article of fairground equipment;

(b) any article designed for use as a component in any such place or equipment.

'Substance' means any natural or artificial substance (including micro-organisms) intended for use (whether exclusively or not) by persons at work.

STATEMENTS OF HEALTH AND SAFETY POLICY

There is a general duty on every employer under section 2 of the HSWA "to prepare and, as often as may be necessary, revise a written statement of his general policy with respect to the health and safety at work of his employees and the organisation and arrangements for the time being in force for carrying out that policy". Moreover, he must bring the statement, and any revision of it, to the notice of all his employees. (Where less than five employees are employed, the statement need not be in writing.)

The principal features of a Statement of Health and Safety Policy are:

(a) a statement of intent, which outlines the organisation's overall philosophy in relation to the management of health and safety, reflecting the duties of employers under section 2 of the HSWA and including the objectives for ensuring legal compliance;

(b) the organisation in respect of health and safety, which should indicate the chain of responsibility and accountability from senior management level downwards;

(c) the arrangements which detail the procedures and systems for monitoring health and safety performance and the overall implementation of the objectives detailed in the statement of intent including, for example, the provision of information, instruction and training, risk assessment procedures and accident reporting, recording and investigation procedures.

Many "Statements of Health and Safety Policy" incorporate a series of secondary policy statements, such as individual policies on smoking at work, health and safety training and health surveillance.

It is also common practice to incorporate appendices in a "Statement of

Health and Safety Policy". These appendices might cover, for instance, a list of the relevant statutory provisions applying to the organisation, individual responsibilities for health and safety and sources of information for employees.

CORPORATE LIABILITY

This is concerned with the duties of the 'body corporate' and its officers. Thus, where an offence under any of the relevant statutory provisions committed by a body corporate is proved to have been committed with the consent or connivance of, or to have been attributable to any neglect on the part of, any director, manager, secretary or other similar officer of the body corporate or a person who was purporting to act in such capacity, he as well as the body corporate shall be guilty of that offence and shall be liable to be proceeded against and punished accordingly (section 37 HSWA).

The implications of section 37 are significant in that:

(a) where an offence is committed through neglect or omission by a constituted board of directors, the organisation itself can be prosecuted together with those directors who, individually, may have been at fault;

(b) where an individual functional director, e.g. a production director, engineering director, is guilty of an offence, he can be prosecuted personally as well as the organisation;

(c) an organisation can be prosecuted even though the act or omission resulting in the offence was committed by a junior official, e.g. a supervisor or foreman, or executive, or even a visitor to the organisation's premises.

Other 'corporate' persons, e.g. chief engineers, training managers, personnel managers, health and safety advisers, may also be liable to prosecution.

Section 36 of the HSWA deals with such offences thus:

> Where the commission by any person of an offence under any of the relevant statutory provisions is due to the act or default of some other person, that other person shall be guilty of the offence and a person may be charged with and convicted of the offence whether or not proceedings are taken against the first mentioned person.

ENFORCEMENT ARRANGEMENTS

The Enforcing Authorities

The enforcing authorities under HSWA are:

(a) the Health and Safety Executive (HSE), which is divided into a number of specific inspectorates e.g. Factories, Nuclear Installations, Construction;

(b) Local Authorities, principally through their environmental health departments; and

(c) for certain matters, the local fire authority.

Enforcement is undertaken by inspectors appointed under the HSWA and authorised by a written warrant from the enforcing authority.

Powers of Inspectors

Under section 20 of the HSWA an inspector has the following powers:

(a) to enter premises at any reasonable time accompanied, if necessary, by police officers;

(b) to take with him any duly authorised person, equipment or materials required;

(c) to make examinations and investigations;

(d) to direct that any premises, any part thereof or anything therein shall remain undisturbed for the purposes of examination and investigation;

(e) to take measurements, photographs, recordings and samples;

(f) to cause any article or substance to be dismantled or subjected to any process or test;

(g) to take possession of any article or substance and to detain for as long as is necessary;

(h) to require any person to give information, answer questions and sign a declaration of truth;

(i) to require production of, inspect and take copies of books and documents required to be maintained or otherwise;

(j) to require any person to afford appropriate facilities and assistance;

(k) to inform safety representatives of matters he has found following an investigation or examination;

(l) to serve *Improvement Notices* and *Prohibition Notices*;

(m) to prosecute offenders in the criminal courts.

NOTICES UNDER THE HSWA

An inspector appointed under the HSWA may serve two types of notice.

Improvement Notices

Where an inspector is of the opinion that a person:

(a) is contravening one or more of the relevant statutory provisions; or

(b) has contravened one or more of these provisions in circumstances that make it likely that the contravention will continue or be repeated.

The inspector may serve an improvement notice on that person requiring that he remedy the contravention(s) or, as the case may be, the matters occasioning it within such period (ending not earlier than the period within which an appeal may be brought) as may be specified in the notice. (*See Figure 1: Specimen Improvement Notice.*)

Note: The 'Relevant Statutory Provisions'

The HSWA is an 'umbrella' Act of general duties over, in some cases, former legislation, such as the Factories Act 1961 and the Offices, Shops and Railway Premises Act 1963. Both Acts, and any Regulations made under them, are deemed to be the relevant statutory provisions with regard to the HSWA.

Schedule 1 to the HSWA lists Regulations deemed to be relevant statutory provisions. Moreover, section 15 of the HSWA gives the Secretary of State power to make Regulations that are part of the relevant statutory provisions.

The relevant statutory provisions include:

(a) Part I of the HSWA;

(b) Regulations made under Part I;

(c) the Acts contained in Schedule I of the HSWA;

(d) any Regulations made under the above Acts.

Prohibition Notices

Where an inspector is of the opinion that a work activity involves or will involve a risk of serious personal injury he may serve a prohibition notice on

Figure 1: Specimen Improvement Notice

HEALTH AND SAFETY EXECUTIVE   Serial No. I

Health and Safety at Work etc. Act 1974, Sections 21, 23 and 24

IMPROVEMENT NOTICE

Name and address (See Section 46)	To ....
(a) Delete as necessary	....
	(a) Trading as ....
(b) Inspector's full name	(b) ....
	one of (c) ....
(c) Inspector's official designation	of (d) ....
	.... Tel No. : ....
(d) Official address	hereby give you notice That I am of the opinion that at ....
(e) Location of premises or place and activity	(e) ....
	you, as (a) an employer/s self employed person/s person wholly or partly in control of the premises
	(f) ....
(f) Other specified capacity	(a) are contravening/have contravened in circumstances that make it likely that the contravention will continue or be repeated
	....
	....
(g) Provisions contravened	(g) ....
	....
	The reasons for my said opinion are:– ....
	....
	....
	and I hereby require you to remedy the said contraventions or, as the case may be, the matters occasioning them by
(h) Date	(h) ....
	(a) In the manner stated in the attached schedule which forms part of the notice.
	Signature .... Date ....
	Being an inspector appointed by an instrument in writing made pursuant to Section 19 of the said Act and entitled to issue this notice.
	(a) An Improvement notice is also being served on
	....
	of ....
LP1	related to the matters contained in this notice.

the owner and/or occupier of the premises or the person having control of that activity. Such a notice will direct that activities specified in the notice shall not be carried on by or under the control of person on whom the notice is served unless certain specified remedial measures have been complied with.

It should be appreciated that it is not necessary that an inspector believe that a legal provision is being or has been contravened. A prohibition notice is served where there is immediate threat to life and in anticipation of danger.

A prohibition notice may be served with immediate effect, or it may be deferred, thereby allowing the recipient of the notice a short period of time, say 48 hours, to remedy the situation, undertake modifications, carry out works, etc. The duration of a deferred prohibition notice is stated on the notice. (*See Figure 2: Specimen Prohibition Notice.*)

Failure to comply with a Notice

With both an improvement notice and a prohibition notice, where:

(a) there is a failure to comply within the time specified; or

(b) in the event of an appeal against the notice, after the expiry of any extra time allowed for compliance by a tribunal,

the person concerned can be prosecuted. Furthermore, where a person is convicted of an offence specified in an improvement or prohibition notice, and the contravention is continued after the conviction, he may be found guilty of a further offence and liable on summary conviction to a fine not exceeding £200 per day for each day on which the contravention is continued, in addition to the original maximum fine of £20,000 for each contravention specified.

Appeals Against Notices

A person served with an improvement or prohibition notice can appeal to a tribunal. In the case of an improvement notice, the lodging of the appeal results in the notice being automatically suspended. However, in the case of a prohibition notice, the notice stays in force, unless a tribunal directs otherwise.

Prosecution

Failure to comply with either notice may result in prosecution. Moreover, an inspector does not necessarily have to serve a notice, but can institute legal proceedings without serving a notice.

Cases are heard in the criminal courts, namely the Magistrates Court and, on indictment, the Crown Court. Much will depend on the seriousness of the offence. (*See Figure 3: Legal Routes following an Accident at Work.*)

Figure 2: Specimen Prohibition Notice

HEALTH AND SAFETY EXECUTIVE

Health and Safety at Work etc. Act 1974, Sections 22–24 Serial No. P

PROHIBITION NOTICE

Name and address (See Section 46)

(a) Delete as necessary

(b) Inspector's full name

(c) Inspector's official designation

(d) Official address

To ....

....

(a) Trading as ....

(b) ....

one of (c) ....

of (d) ....

.... tel no..........................

hereby give you notice that I am of the opinion that the following activities,

namely:– ....

....

....

which are (a) being carried on by you/about to be carried on by you/under your control

(e) Location of activity

at (e) ....

involve, or will involve (a) a risk/an imminent risk, of serious personal injury. I am further of the opinion that the said matters involve contraventions of the following statutory provisions:–......

....

....

....

because ....

....

....

and I hereby direct that the said activities shall not be carried on by you or under your control (a) immediately/after

(f) Date

(f)..

unless the said contraventions and matters included in the schedule, which forms part of this notice, have been remedied.

Signature .... Date ....

being an inspector appointed by an instrument in writing made pursuant to Section 19 of the said Act and entitled to issue this notice.

LP2

Figure 3: Legal Routes following an Accident at Work

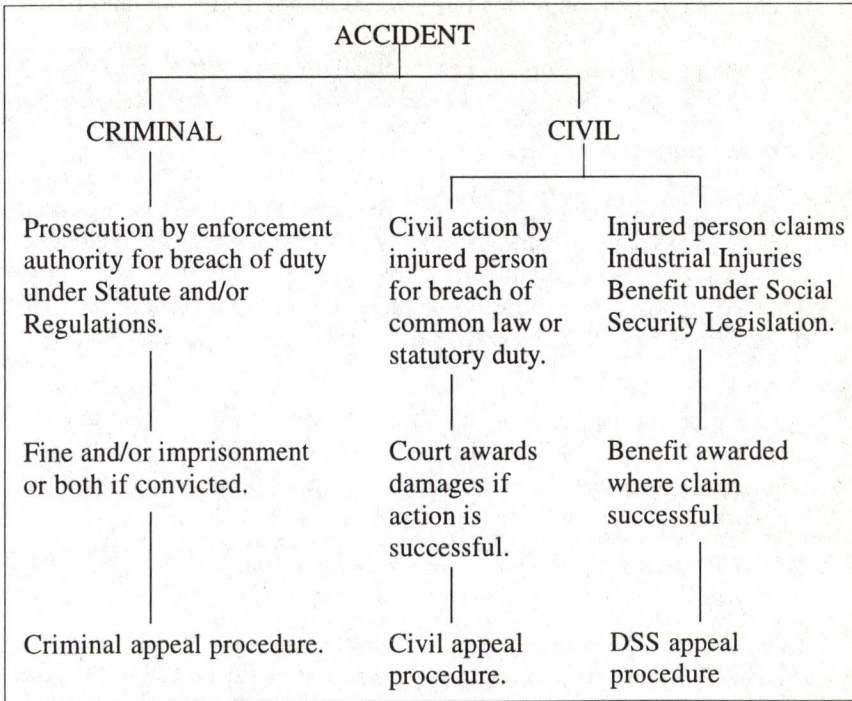

```
                            ACCIDENT
              ┌──────────────────────────────────┐
        CRIMINAL                              CIVIL
           │                        ┌────────────────────┐

Prosecution by enforcement    Civil action by      Injured person claims
authority for breach of duty  injured person       Industrial Injuries
under Statute and/or          for breach of        Benefit under Social
Regulations.                  common law or        Security Legislation.
                              statutory duty.


Fine and/or imprisonment      Court awards         Benefit awarded
or both if convicted.         damages if           where claim
                              action is            successful
                              successful.


Criminal appeal procedure.    Civil appeal         DSS appeal
                              procedure.           procedure
```

MANAGEMENT OF HEALTH AND SAFETY AT WORK
REGULATIONS (MHSWR) 1992

These Regulations, which are accompanied by an ACOP, implemented the European Framework Directive "on the introduction of measures to encourage improvements in the safety and health of workers at work". They are significant in that they brought in the concept of 'risk assessment', which forms the basis for all modern protective legislation. Furthermore, compared with the HSWA, duties on employers and others under the MHSWR of an absolute or strict nature whereas duties under the HSWA are largely qualified by 'so far as is reasonably practicable'. In fact, this increase in the level of duty has been the trend with the majority of recent legislation, such as the Workplace (Health, Safety and Welfare) Regulations 1992, the Construction (Design and Management) Regulations 1994 and the Fire Precautions (Workplaces) Regulations 1997. (*See 'Current trends in Health and Safety Legislation' at the end of this chapter.*)

What must further be appreciated is the fact that the MSWR do not stand on their own. All other Regulations must be read in conjunction with the general duties detailed in the MSWR, e.g. risk assessment, the appointment of competent persons and the provision of information, instruction and training.

The principal requirements of these regulations are outlined below.

Duties of Employers

1. To make a suitable and sufficient assessment of the risks to his own employees and to other persons affected by his activities in order to identify the measures he needs to take to comply with the requirements and prohibitions imposed upon him by or under the relevant statutory provisions. (Similar provisions apply in the case of self-employed persons.)

    Note
    Before a person can undertake a 'suitable and sufficient' risk assessment, he should be fully aware of 'the requirements and prohibitions imposed by or under the relevant statutory provisions' which apply to the workplace, work activity, etc. being assessed.

2. To review and revise risk assessments, and implement any changes, where necessary.

3. To ensure the effective planning, organisation, control, monitoring and review of the preventive and protective measures (identified in the risk assessment).

4. To provide health surveillance for employees where the need is identified in the risk assessment.

5. To appoint one or more competent persons to assist him in complying with the requirements and prohibitions imposed by or under the relevant statutory provisions.

6. To establish and, where necessary, give effect to procedures to be followed in the event of serious or imminent danger, and to nominate competent persons to implement these procedures.

7. To provide employees with comprehensible and relevant information on:
    (a) the risks identified by a risk assessment;
    (b) the preventive and protective measures;
    (c) the emergency procedures in the event of serious or imminent danger;
    (d) the identity of the competent persons generally and those nominated to implement emergency procedures;
    (e) the risks associated with shared workplaces, where appropriate.

8. Where a workplace is shared with other employers (e.g. a construction site, office block, industrial estate):

(a) to co-operate with other employers on legal compliance;
(b) to take all reasonable steps to co-ordinate safety procedures;
(c) to inform other employers of the risks arising out of or in connection with his own undertaking.

9. To provide comprehensible information to employers from an outside undertaking on the risks arising from his own undertaking and the measures he has taken to comply with the relevant statutory provisions.

10. To take into account health and safety-related capabilities of individuals when entrusting tasks to them.

11. To ensure the health and safety training of employees:
(a) at the recruitment stage;
(b) on being exposed to new or increased risks due to transfer, change of responsibility, the introduction of new work equipment, a change respecting existing work equipment, the introduction of new technology, the introduction of a new system of work or any change respecting an existing system of work.

Such training to be repeated periodically, adapted to take account of new or changed risks, and to be undertaken during working hours.

12. To provide temporary workers with comprehensive information on:
(a) any occupational skills or qualifications they may require to ensure safe working;
(b) any health surveillance required in accordance with the requirements of the relevant statutory provisions.

13. To consult with safety representatives as follows:
(a) concerning the introduction of any measure which may substantially affect the health and safety of employees that the representative represents;
(b) about any arrangements for nominating competent persons;
(c) with regard to information he is required to provide to employees in accordance with the relevant statutory provisions;
(d) on the planning and organisation of any health and safety training;
(e) with regard to the consequences of the introduction of new technology.

14. To provide facilities and assistance to enable safety representatives to carry out their functions.

Duties of Employees

There are absolute duties on employees to:

(a) use any machinery, equipment, dangerous substances, transport equipment, means of production or safety device in accordance with any training or instructions received;

(b) report situations of serious or immediate danger, together with any shortcomings in the employer's protection arrangements to their employees.

THE COURT HIERARCHY

There are two distinct systems of courts, namely those courts dealing with criminal matters and those dealing with civil matters. (*See Figure 4: The Structure of the Courts.*)

Magistrates Court

As the 'court of summary jurisdiction' or the 'court of first instance', the magistrates court is the lowest of the courts in England and Wales. It deals principally with criminal matters and its jurisdiction is limited. On this basis, magistrates determine and sentence for the majority of the less serious offences. They also hold preliminary examinations into more serious offences to ascertain whether the prosecution can show a *prima facie* case on which the accused may be committed for trial at a higher court, namely the Crown Court. The Sheriffs Court performs a similar function in Scotland.

Crown Court

Serious criminal charges and cases, where an accused has the right to trial by jury, are heard on indictment in the Crown Court before a judge and jury. This court also hears appeals from Magistrates Courts.

County Court

These courts deal with in the first instance with civil matters, e.g. civil claims in respect of negligence or breach of statutory duty, and have limited jurisdiction. They operate on an area basis and cases are normally heard by circuit judges and registrars.

High Court of Justice

This court deals with more important and complex civil matters, cases being heard by a High Court judge. The High Court has three divisions:

(a) *Queen's Bench* deals with contract and torts, and has a supervisory function over the lower courts and tribunals;

(b) Chancery   deals with matters covering a range of aspects, such as wills, bankruptcy, partnerships and companies;

(c) Family   deals with matters involving, for instance, the adoption of children, marital disputes and property.

In addition, the Queen's Bench Division hears appeals on matters of law:

(a) from the Magistrates Courts and from the Crown Court on a 'case stated' basis;

(b) from some tribunals, e.g. the findings of a tribunal on an enforcement notice under the HSWA.

Court of Appeal

The Court of Appeal has two divisions:

(a) the Civil Division, which hears appeals from the County Courts and the High Court of Justice;

(b) the Criminal Division, which hears appeals from the Crown Court.

The High Court, Crown Court and Court of Appeal are known as the Supreme Court of Judicature.

House of Lords

The House of Lords hears appeals on important legal issues only from the Court of Appeal and, in restricted circumstances, from the High Court. Such issues are dealt with by the Law Lords.

European Court of Justice

This is the supreme law court, whose decisions on the interpretation of European Union law are sacrosanct. Cases can only be brought before this court by organisations or by individuals representing organisations. Decisions are enforceable through the network of courts and tribunals in all Member States.

INDUSTRIAL TRIBUNALS

Tribunals deal with a wide range of industrial issues, including those involving industrial relations, equal pay and sex discrimination. They consist of a legally qualified chairperson (appointed by the Lord Chancellor) and two lay members, one from management and one from a trade union. When all three members of a tribunal are sitting, the majority view prevails.

Health and Safety Issues

Tribunals deal with:

(a) appeals against improvement and prohibition notices;

(b) time off for the training of safety representatives;

(c) failure by an employer to a pay a safety representative for time off whilst undertaking his functions and during training;

(d) dismissal, actual or constructive, following a breach of health and safety law and/or a term of an employment contract;

(e) failure by an employer to make a medical suspension payment. Employment Protection (Consolidation) Act 1978.

Figure 4: Structure of the Courts in England and Wales

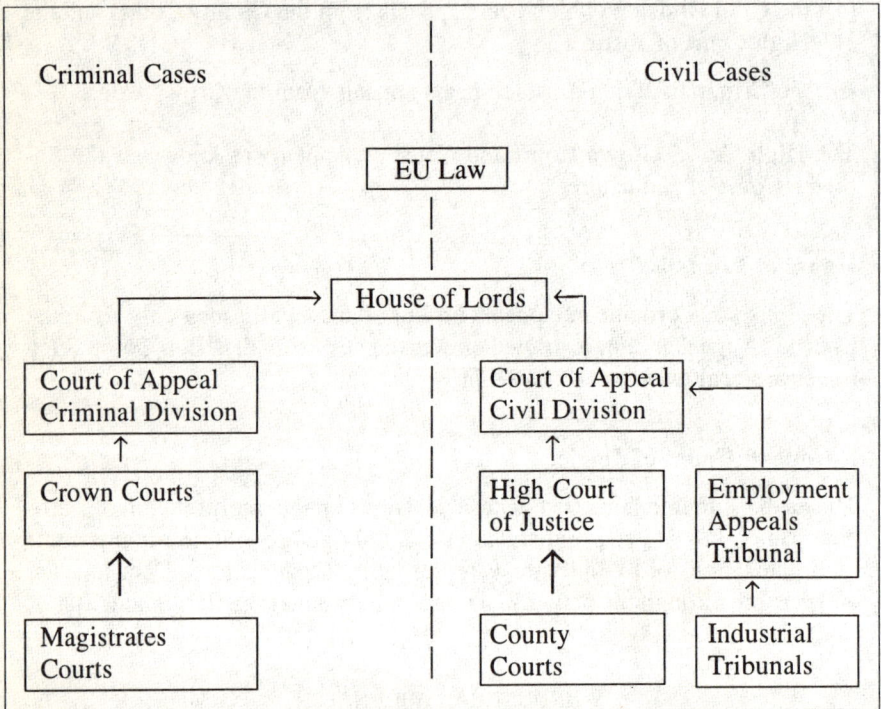

```
Criminal Cases                                            Civil Cases

                          ┌─────────────┐
                          │   EU Law    │
                          └─────────────┘

                     →  House of Lords  ←

┌────────────────────┐              ┌────────────────────┐
│ Court of Appeal    │              │ Court of Appeal    │  ←
│ Criminal Division  │              │ Civil Division     │
└────────────────────┘              └────────────────────┘
          ↑                                   ↑
┌────────────────────┐    ┌──────────────┐  ┌──────────────┐
│ Crown Courts       │    │ High Court   │  │ Employment   │
│                    │    │ of Justice   │  │ Appeals      │
└────────────────────┘    └──────────────┘  │ Tribunal     │
          ↑                      ↑           └──────────────┘
                                                    ↑
┌────────────────────┐    ┌──────────────┐  ┌──────────────┐
│ Magistrates        │    │ County       │  │ Industrial   │
│ Courts             │    │ Courts       │  │ Tribunals    │
└────────────────────┘    └──────────────┘  └──────────────┘
```

JOINT CONSULTATION

Procedures and arrangements for consultation on health and safety issues between employers and employees are covered by specific regulations.

Safety Representatives and Safety Committees Regulations (SRSCR) 1977

These regulations are concerned with the appointment by recognised trade unions of safety representatives, the functions of such persons and the establishment and operation of safety committees. The regulations are accompanied by an ACOP and HSE Guidance.

Safety Representatives

A safety representative is a person appointed by his trade union to represent members of that trade union in consultations with the employer on all matters relating to health and safety at work. He has the following specific functions, following notification to his employer by the trade union, of his appointment:

(a) to investigate potential hazards and causes of accidents at the workplace;

(b) to investigate complaints from employees concerning risks to health and safety at work;

(c) to make representations to the employer on matters arising out of (a) and (b) above and on general matters affecting the health and safety of employees;

(d) to carry out certain inspections:
    (i) of the workplace, after giving reasonable notice to the employer;
    (ii) of a relevant area following a notifiable and reportable accident or scheduled dangerous occurrence, or where a reportable disease is contracted, if it is safe to do so and in the interests of the employees he represents;
    (iii) of documents relating to the workplace or employees which the employer is required to maintain;

(e) to represent his group of employees in consultation with inspectors appointed under the Act, and to receive information from inspectors;

(f) to attend meetings of safety committees.

Safety Committees

Where an employer is requested in writing by at least two safety representatives to do so, he must form a safety committee. Whilst it is the employer's prerogative to establish the objectives, role and function of such a committee, together with the arrangements for running same, he must:

(a) consult with both the safety representatives making this request and with representatives of the trade union whose members work in any workplace where it is proposed that the committee will function;

(b) post a notice stating the composition of the committee, and the workplace(s) to be covered by it, in a place where it can be easily read by employees;

(c) establish the committee within three months following the request for its formation.

Health and Safety (Consultation with Employees) Regulations 1996

These Regulations changed the law with regard to the health and safety consultation process between employers and employees – mainly because many employees are not members of a recognised trade union.

Under the SRSCR, employers must consult safety representatives appointed by any trade unions they recognise. Under the 1996 Regulations employers must consult any employees who are not covered by the SRSCR. This may be by direct consultation with the employees concerned or through representatives elected by the employees they represent.

HSE Guidance accompany these regulations indicates:

(a) which employees must be involved;

(b) the information with which they must be provided;

(c) procedures for the election of representatives of employee safety;

(d) the training, time off and facilities with which they must be provided;

(e) their functions in office.

STANDARD INFORMATION TO EMPLOYEES

The Health and Safety (Information for Employees) Regulations 1989 require information relating to health, safety and welfare to be furnished to employees by means of posters or leaflets in the form approved and published for the purposes of the regulations by the HSE. Copies of the form of poster or leaflets approved in this way may be obtained from government bookshops.

The Regulations also require the name and address of the enforcing authority and the address of the employment medical advisory service to be written in the appropriate space on the poster and, where the leaflet is given out instead, the same information should be specified in a written notice accompanying it. (*This leaflet is reproduced in Figure 5.*)

Figure 5: Information for Employees Leaflet

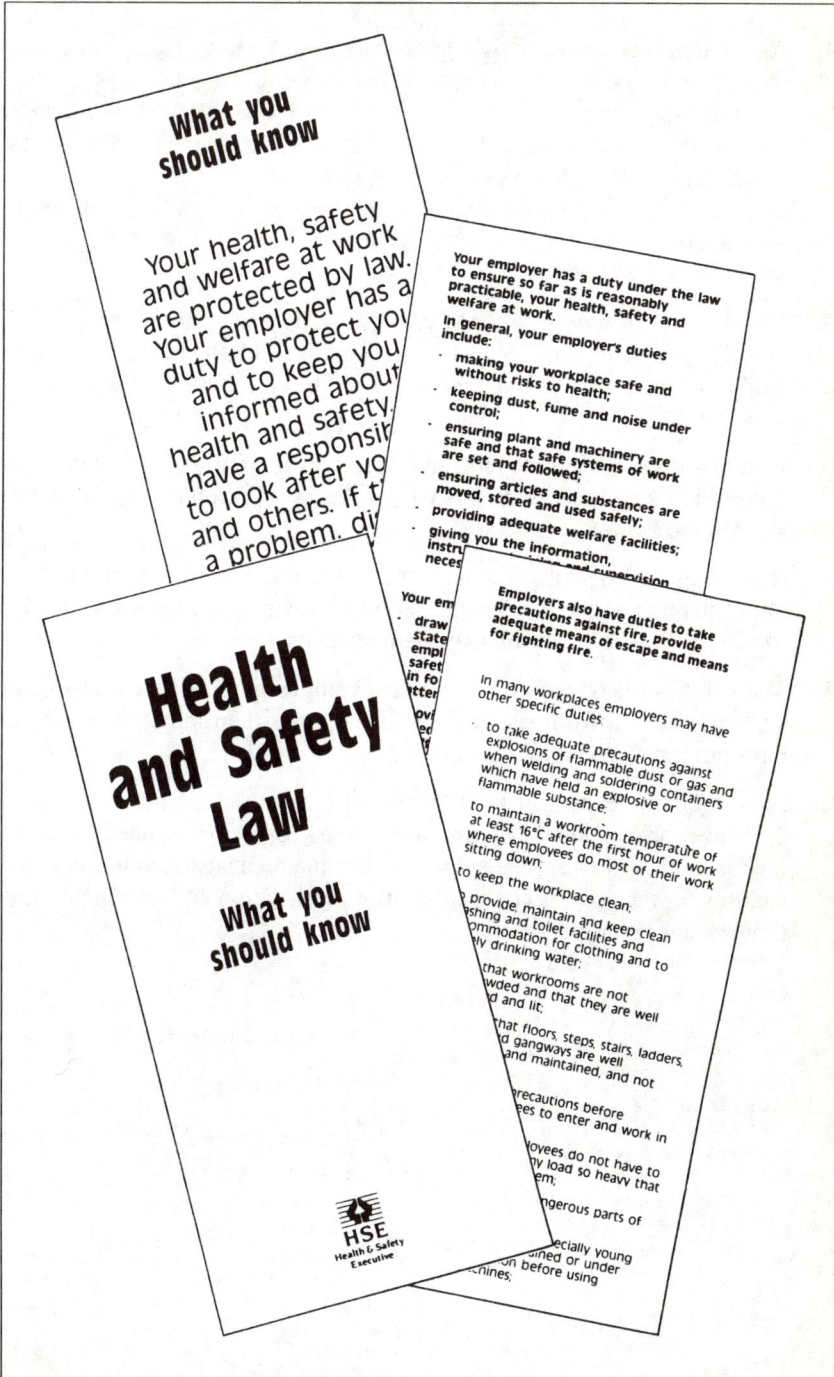

CURRENT TRENDS IN HEALTH AND SAFETY LEGISLATION

1. All modern protective legislation is largely driven by European Direc-
    tives. For instance:
    (a) the Framework Directive on 'the health and safety of workers at
        work' was implemented in the UK as the Management of Health and
        Safety at Work Regulations 1992; and
    (b) the Temporary and Mobile Construction Sites Directive was imple-
        mented in the UK as the Construction (Design and Management)
        Regulations 1994.

2. Regulations brought into force since 1992 do not, in most cases, stand
    on their own. The must be read in conjunction with the general duties on
    employers under the Management of Health and Safety at Work Regula-
    tions 1992.

3. Duties imposed on employers tend largely to be of an absolute nature, as
    opposed to qualified duties, such as 'so far as is reasonably practicable',
    as with the HSWA.

4. Risk assessment, taking into account the 'requirements and prohibitions
    imposed by or under the relevant statutory provisions' is the starting point
    for all health and safety management systems.

5. The emphasis is on the planning and development of safety management
    systems derived from risk assessment, sooner than merely complying
    with minimum prescriptive standards.

6. Most modern health and safety legislation requires, or implies the need
    for, some form of documentation, such as risk assessments, planned safety
    management systems, planned preventive maintenance systems and pro-
    cedures for the provision of information, instruction and training of em-
    ployees and other persons.

SUMMARY – CHAPTER 1

1. Health and safety law is founded on both statute law and common law.

2. The common law enables employees, their dependants and others to sue an employer for damages as a result of death, disease or injury at work.

3. Negligence and/or breach of statutory duty form the basis for most civil claims.

4. Duties under health and safety law may be absolute or qualified.

5. The Secretary of State for employment is empowered to make regulations under the HSWA.

6. The HSWA places duties on employers, employees, occupiers of premises and manufacturers and designers of articles and substances used at work.

7. All employers must produce and revise where necessary a *Statement of Health and Safety Policy*.

8. Enforcing authorities have wide powers under the HSWA.

9. The Management of Health and Safety at Work Regulations 1992 brought in important requirements with regard to risk assessment, management systems, competent persons, the provision of information and training and the consideration of human capability.

10. Both criminal and civil courts deal with health and safety-related issues.

11. Employers should have well-organised procedures for consultation with employees.

CHAPTER 2

HEALTH AND SAFETY MANAGEMENT

RISK ASSESSMENT

Risk assessment is the starting point for most health and safety management systems. A risk assessment may be defined as:

> An identification of the hazards present in an undertaking and an estimate of the extent of the risks involved, taking into account whatever precautions are already being taken.

It is essentially a four-stage process:

(a) the identification of all the hazards;

(b) the measurement of the risks;

(c) the evaluation of the risks;

(d) the implementation of measures to eliminate or control the risks.

Approaches to Risk Assessment

There are different approaches that can be adopted in the workplace:

(a) the examination of each activity which could cause injury;

(b) the examination of hazards and risks in groups, e.g. machinery, substances or transport;

(c) the examination of specific departments, sections, offices, construction sites.

Features of Risk Assessment

A risk assessment should:

(a) identify all the hazards associated with the operation, and evaluate the risks arising from those hazards, taking into account current legal requirements;

(b) record any significant findings;

(c) identify any group of employees, or single employees as the case may be, who are especially at risk;

(d) identify others who may be specially at risk, e.g. visitors, contractors or members of the public;

(e) evaluate existing controls, stating whether or not they are satisfactory and, if not, what action should be taken;

(f) evaluate the need for information, instruction, training and supervision;

(g) judge and record the probability or likelihood of an accident occurring as a result of uncontrolled risk, including the outcome of the 'worst case' likely;

(h) record any circumstances arising from the assessment where serious, imminent and unavoidable danger could arise;

(j) provide an action plan giving information on implementation of additional controls, in order of priority, and with a realistic timescale.

Recording the Assessment

As stated, the assessment must be recorded and must incorporate details of items (a) to (j) above. (Electronic methods of recording are acceptable.)

Generic Assessments

These are assessments produced once only for a given activity or type of workplace. In cases where an organisation has several locations or situations where the same activity is undertaken then a generic risk assessment could be carried out for a specific activity to cover all locations. Similarly, where operators work away from the main location and undertake a specific task, e.g. installation of telephones or servicing of equipment, a generic assessment should be produced.

For generic assessments to be effective:

(a) 'worst case' situations must be considered;

(b) provision should be made in the assessment to monitor the implementation of the assessment controls that are/are not relevant at a particular location and to identify what action needs to be taken in order to implement the relevant, required actions from the assessment.

In certain cases, there may be risks that are specific to one situation only, and these may need to be incorporated in a separate part of the generic risk assessment.

Maintaining the Risk Assessment

The risk assessment must be maintained. This means that any significant change to a workplace, process or activity, or the introduction of any new process, activity or operation, should be subject to risk assessment. If new hazards come to light, then these should also be subject to risk assessment.

The risk assessment, furthermore, should be periodically reviewed and updated, this is best achieved by a suitable combination of safety inspection and monitoring techniques, which require corrective and/or additional action where the need is identified.

Typical monitoring systems include:

(a) preventive maintenance inspections;

(b) safety representative/committee inspections;

(c) statutory and maintenance scheme inspections, tests and examinations;

(d) safety tours and inspections;

(e) occupational health surveys;

(f) air monitoring;

(g) safety audits.

Useful information on checking performance against control standards can also be obtained reactively from the following activities:

(a) by the investigation of accidents and ill-health;

(b) by the investigation of any damage to plant, equipment and vehicles;

(c) by the investigation of 'near miss' situations.

Reviewing the Risk Assessment

The frequency of review depends upon the level of risk but this frequency should be stated in the risk assessment document. Further, if a serious accident occurs in the organisation, or elsewhere but is possible in the organisation, and where a check on the risk assessment shows no assessment or a gap in assessment procedures, then a review is necessary.

Risk/Hazard Control

Once the risk or hazard has been identified and assessed, employers must either prevent the risk arising or, alternatively, control it. Much will depend upon the magnitude of the risk in terms of the controls applied. In certain cases, the level of competence of operators may need to be assessed prior to their undertaking certain work, e.g. work on electrical systems.

A typical hierarchy of control, from high risk to low risk, is indicated below.

1. *Elimination* of the risk completely, e.g. prohibiting a certain practice or the use of a certain hazardous substance.

2. *Substitution* by something less hazardous or risky.

3. *Enclosure* of the risk in such a way that access is denied.

4. *Guarding* or the installation of safety devices to prevent access to danger points or zones on work equipment and machinery.

5. *Safe systems of work* that reduce the risk to an acceptable level.

6. *Written procedures*, e.g. job safety instructions, that are known and understood by those affected.

7. *Adequate supervision*, particularly in the case of young or inexperienced persons.

8. *Training of staff* to appreciate the risks and hazards.

9. *Information*, e.g. safety signs, warning notices.

10. *Personal protective equipment*, e.g. eye, hand, head and other forms of body protection.

In many cases, a combination of the above control methods may be necessary. It should be appreciated that the amount of management control necessary will increase proportionately for the controls lower down this list.

The Principal Types of Hazard

It may be necessary to consider the following hazards when undertaking risk assessments.

Fall of person from a height	Electricity
Fall of an object/material from a height	Drowning
Fall of a person on the same level	Excavation work
Manual handling	Stored energy
Use of work equipment	Explosions
Operation of vehicles	Contact with hot/cold surfaces
Compressed air	Adverse weather
Mechanical lifting operations	Hazardous substances
Noise and vibration	Storage of goods
Biological agents	Housekeeping/cleaning
Radiation	Temperature, lighting and
Fire	ventilation

HSE Guidance on Risk Assessment

The HSE leaflet *Five Steps to Risk Assessment* (INDG163.rev 1.1998) provides guidance to firms in the commercial, service and light industrial sectors. The principal elements of this leaflet are outlined in figure 6 below:

Figure 6: Five Steps to Risk Assessment

How to assess the risks in your workplace
Follow the five steps in this leaflet:

STEP 1   Look for the hazards.

STEP 2   Decide who might be harmed and how.

STEP 3   Evaluate the risks and decide whether the existing precautions are adequate or whether more should be done.

STEP 4   Record your findings.

STEP 5   Review your assessment and revise it if necessary.

STEP 1: Hazard

Hazard means anything that can cause harm (e.g. chemicals, electricity, working from ladders, etc.).

Look only for hazards which you could reasonably expect to result in significant harm under the conditions in your workplace. Use the following examples as a guide.

- slipping/tripping hazards   e.g. poorly maintained floors or stairs;
- fire   e.g. from flammable materials;
- chemicals   e.g. battery acid;
- work at height   e.g. from mezzanine floors;
- ejection of material   e.g. from plastic moulding;
- pressure systems   e.g. steam boilers;
- vehicles   e.g. fork-lift trucks;
- electricity   e.g. poor wiring;
- dust   e.g. from grinding;
- fumes   e.g. from welding;
- manual handling;
- moving parts of machinery;
- noise;
- poor lighting;
- low temperature.

STEP 2: Who might be harmed?

There is no need to list individuals by name – just think about the groups of people doing similar work or who may be affected. For example:

- office staff;
- maintenance personnel;
- contractors;
- people sharing your workplace;
- operators;
- cleaners;
- members of the public.

Pay particular attention to:

- staff with disabilities;
- visitors;
- inexperienced staff;
- lone workers.

They may be more vulnerable.

STEP 3: Is more needed to control the risk?

Risk is the chance, high or low, that somebody will be harmed by the hazard. For the hazards listed, do the precautions already taken:

- meet the standards set by a legal requirement?
- comply with a recognised industry standard?
- represent good practice?
- reduce risk as far as is reasonably practicable?

Have you provided:

- adequate information, instruction or training?
- adequate systems or procedures?

If so, then the risks are adequately controlled, but you need to indicate the precautions you have in place. (You may refer to procedure, company rules, etc.) Where the risk is not adequately controlled, indicate what more you need to do (the 'action list').

RISK ASSESSMENT FOR	ASSESSMENT UNDERTAKEN	ASSESSMENT REVIEW
Company name	(date)	
Company address	Signed	Date
	Date	
Postcode		

STEP 1	STEP 2	STEP 3
List significant hazards here	List groups of people who are at risk from the significant hazards you have identified	List existing controls or note where the information may be found. List risks which are not adequately controlled and the action needed.

STEP 4: Record your findings

(See specimen Risk Assessment Form)

STEP 5: Review and revision

Set a date for review of the assessment. On review check that the precautions for each hazard still adequately control the risk. Note the outcome. If necessary complete a new page for your risk assessment.

Making changes in your workplace, e.g. when bringing in new machines, substances or procedures may introduce significant new hazards. Look for them and follow the five steps.

JOINT CONSULTATION

Employers have a duty to consult with their employees on matters relating to health, safety and welfare. This may best be undertaken through a safety committee.

Running a Safety Committee

As with any committee, it is essential that the constitution of a safety committee be in written form. The following aspects should be considered when establishing and running a safety committee.

Objectives

To monitor and review the general working arrangements for health and safety and to act as a focus for joint consultation between employer and employees in the prevention of accidents, incidents and occupational ill health.

Composition

The composition of the committee should be determined by local management, but should normally include representation of management and employees, ensuring all functional groups are represented. Other persons may be co-opted to attend specific meetings, e.g. health and safety adviser or company engineer.

Election of Committee Members

The Chairman, the Deputy Chairman and the Secretary should be elected for a period of one year.

Nominations for these posts should be submitted by a committee member to the Secretary for inclusion in the agenda of the final meeting in each yearly period. Members elected to office may be re-nominated or re-elected to serve for further terms. Elections should be by ballot and should take place at the last meeting in each yearly period.

Frequency of Meetings

Meetings should be held on a quarterly basis or according to local needs. In exceptional circumstances, extraordinary meetings may be held by agreement of the Chairman.

Agenda and Minutes

The agenda should be circulated to all members at least one week before each Committee meeting. The agenda should include:

(a) **apologies for absence:** members unable to attend a meeting should notify the Secretary and make arrangements for a deputy to attend on their behalf;

(b) **minutes of the previous meeting:** minutes of the meeting should be circulated as widely and as soon after the meeting as possible. All members of the Committee, senior managers, supervisors and trade union representatives should receive personal copies. Additional copies should be posted on notice boards;

(c) **matters arising:** the minutes of each meeting should incorporate an 'Action Column' in which persons identified as having future action to take, as a result of the Committee's decisions, are named.

The named person should submit a written report to the Secretary, which should be read out at the meeting and included in the minutes;

(d) **new items:** items for inclusion in the agenda should be submitted to the Secretary in writing, at least seven days before the meeting. The person requesting the item for inclusion in the agenda should state in writing what action has already been taken through the normal channels of communication. The Chairman will not normally accept items that have not been pursued through these normal channels prior to submission to the Secretary. Items requested for inclusion after the publication of the agenda should be dealt with, at the discretion of the Chairman, as emergency items;

(e) **Safety Adviser's report:** the Safety Adviser should submit a written report to the Committee, copies of which should be issued to each member at least two days prior to the meeting and attached to the minutes.

The Safety Adviser's report should include, for example:

(i) a description of all reportable injuries, diseases and dangerous occurrences that have occurred since the last meeting, together with details of remedial action taken;

(ii) details of any new health and safety legislation directly or indirectly affecting the organisation, together with details of any action that may be necessary;

(iii) information on the outcome of any safety monitoring activities undertaken during the month, e.g. safety inspections of specific areas;

(iv) any other matters which, in the opinion of the Secretary and himself, need a decision from the Committee;

(f) **date, time and place of the next meeting.**

HEALTH AND SAFETY TRAINING

Health and safety training, as with other areas of training, should take place in a number of clearly defined stages.

Identification of Training Needs

A training need is said to exist when the optimum solution to an organisation's problem is through some form of training. For training to be effective, it must be integrated to some extent with the selection and placement policies of the organisation.

Selection procedures must, for instance, ensure that the trainees are capable of learning what is to be taught.

Training needs should be assessed to cover:

(a) induction training for new recruits;

(b) orientation training of existing employees on, for example, promotion, change of job, their exposure to new or increased risks, appointment as competent persons, the introduction of new plant, equipment and technology, and prior to the introduction of safe systems of work;

(c) refresher training directed at maintaining competence.

Development of Training Plan and Programme

Training programmes must be co-ordinated with the current personnel needs of the organisation. The first step in the development of a training programme is that of defining the training objectives. Such objectives or aims may best. be designed by job specification in the case of new training, or by detailed task analysis and job safety analysis in respect of existing jobs.

Implementation of the Training Plan and Programme

Decisions must be made as to the extent of both active and passive learning systems to be incorporated in the programme. Examples of active learning systems are group discussion, role-play, syndicate exercises, programmed learning and field exercises, e.g. safety inspections and audits. Active learning methods reinforce what has already been taught on a passive basis.

Passive learning systems incorporate lectures and the use of visual material, such as films and videos. With a passive learning system the basic objective must be that of imparting knowledge. The principal advantage of passive learning systems is that they provide frameworks and can be used where large numbers of trainees are involved. It should be incorporated as an initial introduction to a subject in particular, and should include rules and procedures, providing they are clear and understandable.

Active learning systems are the most effective form of training once the basic framework is established and there is plenty of time available in the training programme. It is suitable for a subject where there are no 100 per cent correct answers. There is more chance of bringing about attitude change on the part of trainees and the level of interest of trainees is maintained.

Evaluation of the Results

There are two questions that need to be asked at this stage:

(a) have the training objectives been met?

(b) if they have been met, could they have been met more effectively?

Operator training in most industries will need an appraisal of the skills necessary to perform a given task satisfactorily, i.e. efficiently and safely. It is normal, therefore, to incorporate the results of such an appraisal in the basic training objectives.

A further objective of, particularly, health and safety training is to bring about long-term changes in attitude on the part of trainees, which must be linked with job performance. Any decision, therefore, as to whether training objectives have been met, cannot be taken immediately the trainee returns to work or after only a short period of time. It may be several months or even years before a valid evaluation can be made after continuous assessment of the trainee.

The answer to the second question can only be achieved through *feedback* from personnel monitoring the performance of trainees, and from the trainees themselves. This feedback can usefully be employed in setting objectives for further training, in the revision of training content and in the analysis of training needs for all groups within the organisation.

PROVIDING HEALTH AND SAFETY TRAINING

An important feature of the jobs of health and safety specialists, supervisors and line managers is that of preparing and undertaking short training sessions for their staff on general and specific health and safety issues. (See Figure 7 below.)

The following aspects need consideration if such activities are to be successful and get the appropriate messages over to staff:

(a) a list of topics to be covered – e.g. safe systems of work or manual-handling procedures – should be developed, followed by the formulation of a specific programme;

(b) sessions should last no longer than 30 minutes;

(c) extensive use should be made of visual aids – films, videos, slides, flip charts, etc;

(d) topics should, as far as possible, be of direct relevance to the group;

(e) participation should be encouraged with a view to identifying possible misunderstandings or concerns that people have. This is particularly important when introducing a new safe system of work or operating procedure;

(f) topics should be presented in a relatively simple fashion in the language that operators can understand. The use of unfamiliar technical, legal or scientific terminology should be avoided, unless an explanation of such terms is incorporated in the session;

(g) consideration must be given to eliminating any boredom, loss of interest or adverse response on the part of participants. On this basis, talks should be given on as friendly a basis as possible and in a relatively informal atmosphere. Many people respond adversely to a formal classroom situation commonly encountered in staff training activities.

STATEMENTS OF HEALTH AND SAFETY POLICY

Under section 2(3) of HSWA every employer has a duty to "prepare and as often as appropriate revise a written statement of his general policy with respect to the health and safety at work of his employees and the organisation and arrangements for the time being in force for carrying out that policy and to bring the statement and any revision of it to the notice of his employees".

This *Statement of Health and Safety Policy* is, fundamentally, the key document for detailing the management systems and procedures to ensure sound levels of health and safety performance. It should be revised at regular intervals, prior to, particularly, changes in the structure of the organisation, the introduction of new articles and substances, and changes in legal requirements affecting the organisation. Fundamentally, the statement must be seen as a 'living document' which reflects the current organisational arrangements, the hazards and precautions necessary, the individual responsibilities of people and systems for monitoring performance.

Figure 7: Presentation and the Presenter: some important points

---

1. The Presenter

Gestures and mannerisms	Distracting habits
Appearance	Pitch, pace and pause
Body movements	Timing
Eye contact	Punctuality
Firmness	Use of notes and visual aids
Empathy	Conviction, confidence and sincerity

SMILE OCCASIONALLY!

2. The Presentation

Preparation and rehearsal	Construction – beginning, middle
Content	and end
Word pictures	Vocabulary

3. The Environment

Room arrangements	Temperature, lighting and ventilation

4. Support Material

Visuals	Equipment working
Handout material	

Objectives of the 'Statement of Health and Safety Policy'

1. It should affirm long range purpose.

2. It should commit management at all levels and reinforce this purpose in the decision-making process.

3. It should indicate the scope left for decision-making by junior managers.

Scope of the 'Statement of Health and Safety Policy'

A well-written statement should cover the following aspects:

1. Management intent.

2. The 'arrangements' for implementing the policy.

3. Individual accountabilities of directors, line managers, employees and other groups, e.g. contractors.

4. Details of the organisation with respect to both line and staff functions.

5. The role and function of health and safety specialists, e.g. health and safety advisors, nursing advisors, occupational hygienists, occupational health nurses, unit safety officers, occupational physicians, company doctors and trade union appointed safety representatives.

Principal Features of a 'Statement of Health and Safety Policy'

A statement should incorporate the following features.

1. A general statement of intent that states the basic objectives, supplemented by details of the organisation and arrangements (rules and procedures).

2. Definition of both the duties and extent of responsibility at specified line management levels for health and safety, with identification made at the highest level of the individual with overall responsibility for health and safety.

3. Definition of the function of the safety adviser/officer and his relationship to senior and line management made clear.

4. The system for monitoring safety performance and publishing of information about that performance.

5. An identification and analysis of hazards together with the precautions necessary on the part of staff, visitors, contractors, etc.

6. An information system that will be sufficient to produce an identification of needs and that can be used as an indicator of effectiveness of the policy.

7. A policy of the provision of information, instruction and training for all levels of the organisation.

8. A commitment to consultation on health and safety and to a positive form of worker involvement.

9. The statement should bear the signature of the person with ultimate responsibility and accountability for health and safety at work, e.g. Chief Executive, Managing Director etc.

INTERNAL SAFETY CODES AND RULES

The employer has a duty under the HSWA to ensure the provision of information, instruction and training, and to identify the hazards and precautions necessary on the part of staff and other persons, e.g. the employees of contractors.

Inadequate and/or ineffective internal codes of practice, rules and instructions to operators contribute to the causes of accidents and ill health at work. In some cases, rules and instructions may be ambiguous, badly worded or simply not available to the people to whom they are directed.

Codes of Practice

Codes of Practice should, preferably, be incorporated in an organisation's 'Health and Safety Handbook' which provides guidance to managers and staff on various matters, for example:

(a) accident reporting, recording and investigation procedures;

(b) the use of hazardous substances;

(c) the selection and use of personal protective equipment.

Safety Rules

Rules relating to safe working procedures and practices are best covered by a 'Staff Health and Safety Handbook' which is issued to all staff at the induction training stage. The handbook should be written in comprehensible style and extensively illustrated. It should incorporate, for instance, staff responsibilities under health and safety law, safe working practices, the principal hazards they could encounter, the procedure for reporting hazards and accidents to line managers, basic 'DOs' and 'DO NOTs' with regard to safe working and arrangements for welfare.

DEVELOPING A SAFETY CULTURE

'Culture' is variously defined as "a state of manners, taste and intellectual development at a time or place" and "refinement or improvement of mind, tastes, etc. by education and training".

All organisations incorporate a set of cultures that have developed over a period of time. They are associated with the accepted standards of behaviour within that organisation and the development of a specific culture with regard to, for example, quality, customer service and written communication, is a continuing quest for many organisations.

Establishing a Safety Culture: the principles involved

With the greater emphasis on health and safety management implied in the recent legislation, attention should be paid by managers to the establishment and development of the correct safety culture within the organisation. The HSE and CBI have provided guidance on this issue.

JR Rimington reported [The Offshore Safety Regime, HSE Director General's Submission to the Piper Alpha Enquiry (1989)] that the main principles involved in the establishment of a safety culture, which are accepted and observed generally, are:

(a) the acceptance of responsibility at, and from, the top, exercised through a clear chain of command, seen to be actual and felt through the organisation;

(b) a conviction that high standards are achievable through proper management;

(c) setting and monitoring of relevant objectives/targets, based upon satisfactory internal information systems;

(d) systematic identification and assessment of hazards and the devising and exercise of preventive systems which are subject to audit and review; in such approaches, particular attention is given to the investigation of error;

(e) immediate rectification of deficiencies;

(f) promotion and reward of enthusiasm and good results.

Developing a Safety Culture: essential features

In *Developing a Safety Culture* (1991), the CBI maintained that:

> Several features can be identified from the study that are essential to a sound safety culture. A company wishing to improve its performance will need to judge its existing practices against them.
>
> 1. Leadership and commitment from the top which is genuine and visible. This is the most important feature.
>
> 2. Acceptance that it is a long-term strategy which requires sustained effort and interest.
>
> 3. A policy statement of high expectations and conveying a sense of optimism about what is possible supported by adequate codes of practice and safety standards.
>
> 4. Health and safety should be treated as other corporate aims, and properly resourced.
>
> 5. It must be a line management responsibility.
>
> 6. 'Ownership' of health and safety must permeate at all levels of the work force. This requires employee involvement, training and communication.
>
> 7. Realistic and achievable targets should be set and performance measured against them.

8.   Incidents should be thoroughly investigated.

9.   Consistency of behaviour against agreed standards should be achieved by auditing and good safety behaviour should be a condition of employment.

10. Deficiencies revealed by an investigation or audit should be remedied promptly.

11. Management must receive adequate and up-to-date information to be able to assess performance.

Figure 8: DU PONT: Ten Principles of Safety

1.   All injuries and occupational illnesses can be prevented.

2.   Management is directly responsible for preventing injuries and illness, with each level accountable to the one above and responsible for the level below the Chairman undertakes the role of Chief Safety Officer.

3.   Safety is a condition of employment: each employee must assume responsibility for working safely. Safety is as important as production, quality and cost control.

4.   Training is an essential element for safe workplaces. Safety awareness does not come naturally: management must teach, motivate and sustain employee safety knowledge to eliminate injuries.

5.   Safety audits must be conducted. Management must audit performance in the workplace.

6.   All deficiencies must be corrected promptly either through modifying facilities, changing procedures, better employee training or disciplining constructively and consistently. Follow-up audits are use to verify effectiveness.

7.   It is essential to investigate all unsafe practices and incidents with injury potential, as well as injuries.

8.   Safety off the job is as important as safety on the job.

9.   It's good business to prevent illnesses and injuries. They involve tremendous costs. direct and indirect. The highest cost is human suffering.

10. People are the most critical element in the success of a safety and health programme. Management responsibility must be complemented by employees' suggestions and their active involvement.

SUMMARY – CHAPTER 2

1. Risk assessment is the starting point for most health and safety management systems.

2. Risk assessments must, generally, be in writing.

3. A risk assessment should identify the preventive and protective measures to be implemented.

4. Risk assessments should be reviewed on a regular basis.

5. Employers have a duty to consult with employees on health and safety-related issues.

6. Consultation may be through a safety committee, but management retains the responsibility for managing health and safety at the place of work.

7. The provision of information, instruction and training is an important feature of health and safety management.

8. A *Statement of Health and Safety Policy* must incorporate a number of important features.

9. Internal safety codes and rules are an important part of ensuring safe working procedures.

10. Organisations should endeavour to promote and develop an appropriate safety culture.

CHAPTER 3

PRINCIPLES OF ACCIDENT PREVENTION

WHAT IS AN ACCIDENT?

"An unforeseeable event often resulting in injury."

Oxford Dictionary

"A management error; the result of errors or omissions on the part of management."

British Safety Council

"Any deviation from the normal, the expected or the planned usually resulting in injury."

Royal Society for the Prevention of Accidents (RoSPA)

"An unintended or unplanned happening that may or may not result in personal injury, property damage, work process stoppage or interference, or any combination of these conditions under such circumstances that personal injury might have resulted."

Frank Bird, American exponent of 'Total Loss Control'

"An unexpected, unplanned event in a sequence of events that occurs through a combination of causes. It results in physical harm (injury or disease) to an individual, damage to property, business interruption or any combination of these effects."

Health and Safety Unit, University of Aston

THE PRE-ACCIDENT SITUATION

In any situation prior to an accident taking place, two important factors must be considered.

(a) **The objective danger.** This is the objective danger associated with a particular machine, system of work, hazardous substance, etc. at a particular point in time.

(b) **The subjective perception of risk on the part of the individual.** People perceive risks differently according to a number of behavioural factors, such as attitude, motivation, training, visual perception, personality, level of arousal and memory. People also make mistakes. Ergonomic design is significant in preventing human error.

The principal objectives of any accident prevention programme should be, firstly, that of reducing the objective danger present through, for instance, effective standards of machinery safety and, secondly, bringing about an increase in people's perception of risk, through training, supervision and operation of safe systems of work.

HEALTH AND SAFETY STRATEGIES

Pre-Accident Strategies

These can be classified as *Safe Place* and *Safe Person* strategies.

Safe Place Strategies

The principal objective of a safe place strategy is that of bringing about a reduction in the objective danger to people at work. These strategies feature in much of the occupational health and safety legislation that has been enacted over the last century, in particular, the HSWA. Safe place strategies may be classified in the following eight ways.

Safe premises   This relates to the general structural requirements of workplaces, such as the stability of buildings, soundness of floors and the load-bearing capacity of beams. Environmental working conditions, such as the levels of lighting, ventilation and humidity, feature in this classification.

Safe plant   A wide range of plant and machinery and other forms of work equipment, their power sources, location and use are relevant in this case. The safety aspects of individual processes, procedures for vetting and testing new machinery and plant, and systems for maintenance and cleaning must be considered.

Safe processes   All factors contributing to the operation of a specific process must be considered, e.g. work equipment, raw materials, procedures for loading and unloading, the ergonomic aspects of machine operation, hazardous substances used in the process, and the operation of internal workplace transport such as fork lift trucks.

Safe materials   Significant in this case are the health and safety aspects of potentially hazardous chemical substances, radioactive substances, raw materials of all types and specific hazards associated with the handling of these materials. Adequate and suitable information on their correct use, storage and disposal must be provided by manufacturers and suppliers and there may be a need to assess potential health risks.

Safe systems of work   A safe system of work has been defined as "the integration of people, machinery and materials in a correct working environment to provide the safest possible working conditions". The design and implementation of safe systems of work is, perhaps, the most important safe place strategy. It incorporates planning, involvement of operators, training and designing out hazards which may have existed with previous systems of work.

Safe access to and egress from work   This refers to access to, and egress from, both the workplace from the road outside and the working position which may be some distance from the ground, as with construction workers, or several miles below ground in the case of miners. Consideration must be given, therefore, to workplace approach roads, yards, work at high level, the use of portable and fixed access equipment, e.g. ladders and lifts, and the shoring of underground workings.

Adequate supervision   The HSWA requires that in all organisations there must be adequate safety supervision directed by senior management through supervisory management to workers.

Competent and trained personnel   The general duty to train staff and others is also laid down in the HSWA.

All employees need some form of health and safety training. This should be undertaken through induction training and on being exposed to new risks through a change of responsibilities, the introduction of new equipment, the introduction of new technology or new systems of work. Managers must appreciate that a well-trained labour force is a safe labour force and in organisations which carry out health and safety training accident and ill-health costs tend to be lower.

Safe Person Strategies

Generally, safe place strategies provide better protection than safe person strategies. However, where it may not be possible to operate a safe place strategy, then a safe person strategy must be used. In certain cases, a combination of safe place and safe person strategies may be appropriate.

   The main aim of a safe person strategy is to increase people's perception of risk. One of the principal problems of such strategies is they that depend upon the individual conforming to certain prescribed standards and practices, such as the use of certain items of personal protective equipment. Control of the risk is, therefore, placed in the hands of the person whose appreciation of the risk may be lacking or even non-existent. Safe person strategies may be classified in the following five ways.

Care of the vulnerable   In any work situation there will be some people who are more vulnerable to certain risks than others, e.g. where such workers may be exposed to toxic substances, to small levels of radiation or to dangerous metals, such as lead. Typical examples of 'vulnerable' groups are: young people (who through their lack of experience, may be unaware of hazards), pregnant women (where there may be a specific risk to the unborn child) and disabled persons (whose capacity to undertake certain tasks may be limited). In a number of cases there may be a need for continuing medical and/or health surveillance of such persons.

Personal hygiene   The risk of occupational skin conditions caused by contact with hazardous substances, such as solvents, glues, adhesives and a wide range of chemical skin sensitizers, needs consideration. There may also be the risk of ingestion of hazardous substances as a result of contamination of food and drink and their containers. Personal hygiene is very much a matter of individual upbringing. In order to promote good standards of personal hygiene, therefore, it is vital that the organisation provides adequate washing facilities – wash basins, showers, hot and cold water, soap, nailbrushes and drying facilities – for use by workers, particularly prior to the consumption of food and drink and to returning home at the end of the work period.

Personal protective equipment   Generally, the provision and use of any item of personal protective equipment (PPE) must be seen either as a last resort when all other methods of protection have failed or an interim method of protection until some form of safe place strategy can be put into operation. It is by no means a perfect form of protection in that it requires the person at risk to use or wear the equipment all the time they are exposed to a particular hazard. People do not always do this!

Safe behaviour   Employees must not be allowed to indulge in unsafe behaviour or 'horseplay'. Examples of unsafe behaviour include the removal, or defeating the purpose, of machinery guards and safety devices, smoking in designated 'No Smoking' areas, the dangerous driving of vehicles and the failure to wear or use certain items of PPE.

Caution towards danger   All workers and management should appreciate the risks in the workplace, and these risks should be clearly identified in the safety policy required under the HSWA, together with the precautions required to be taken by workers to protect themselves from such risks.

Post-Accident (reactive) Strategies

Whilst principal efforts must go into the implementation of proactive strategies, it is generally accepted that there will always be a need for reactive or post-accident strategies, particularly as a result of failure of the various safe person strategies. The problem with people is that they forget, they take short cuts to save time and effort, they sometimes do not pay attention or they may consider themselves too experienced and skilled to bother about taking basic precautions.

Post-accident strategies can be classified as follows.

Disaster, Contingency, Emergency Planning

Here there is a need for managers to ask themselves this question: "What is the very worst possible type of incident or event that could arise in the business activity?" For most types of organisation this could be a major escalating fire, but other types of major incident should be considered, such as an explosion, collapse of a scaffold, flood or major vehicular traffic accident. The need for some form of emergency plan should be considered.

Feedback Strategies

Accident and ill-health reporting, recording and investigation provides feedback as to the indirect and direct causes of accidents. The study of past accident causes provides information for the development of future proactive strategies. The limitations of accident data as a measure of safety performance should be appreciated.

Improvement, Ameliorative Strategies

These strategies are concerned with minimizing the effects of injuries as quickly as possible following an accident. They will include the provision and maintenance of first aid services, procedures for the rapid hospitalization of injured persons and, possibly, a scheme for rehabilitation following major injury.

THE CAUSES OF ACCIDENTS

The actual causes of accidents are many and varied. In endeavouring to identify the causes of accidents, the following factors should be considered.

Design and layout of working area   Sequential flow aspects; elimination of congestion; storage arrangements; clear gangways and passages; means of escape in the event of fire.

Structural features   Structural safety features of floors; staircases; elevated working platforms. .

Environmental factors   Stress associated with poor standards of control over temperature; lighting; ventilation; humidity; noise and vibration; airborne contaminants.

Operational methods   Correct sequence of operations; adequate space for safe job operation; safe systems of work; use of job safety analysis.

Mechanical or materials failure   Planned inspection of plant; machinery and raw materials.

Maintenance arrangements   Workplace and work equipment maintained "in efficient state, in efficient working order and in good repair"; planned preventive maintenance systems; allocation of responsibilities.

Machinery safety   Machinery hazards; contact with moving parts; items emitted from machines; training and supervision; design and maintenance of machinery guards and safety devices; planned inspections.

Personal protective equipment (PPE)   PPE not provided and/or used; need to assess individual requirements; faulty design; unsuitable equipment, need to assess efficiency of PPE, e.g. respiratory protection; supervision and control.

Cleaning and housekeeping   Poor housekeeping; marking of storage areas; cleaning schedules; allocation of responsibility for cleaning; use of mechanised cleaning equipment.

Supervision   Inadequate supervision; responsibilities of supervisory staff.

Training   Inadequate training; use of untrained staff.

Rules and instructions   Inadequate and/or ambiguous rules and instructions; company *Codes of Practice*; briefing of operators; operator handbooks; breaches of rules and instructions; vigilance by supervisors; reinforcement/ refresher training; disciplinary procedures.

Unsafe attitudes   Unsafe behaviour, horseplay, etc; disciplinary procedures.

Physical and mental disability/incapacity   Need to assess limitations of disabled persons; assessment of human capability prior to allocating tasks.

Ergonomic factors   Aspects of man/machine interface; controls and displays.

Stress   Physical, chemical, biological and psychological stressors at work; changes in work patterns; effects of management decision-making; recognition of the problem of stress at all levels.

Monitoring systems   Inadequate environmental monitoring; hazardous substances; airborne contaminants.

PREVENTION OR CONTROL OF RISKS

In many cases an accident represents a failure on the part of the individual, or his manager/supervisor, to identify and assess the risks associated with a particular workplace, work activity or process. An employer must undertake an assessment of the risks not only to his own employees, but to other persons who may be affected by these risks, e.g. contractors, visitors, members of the public. Once the risks have been assessed, exposure to same must either be prevented or adequately controlled.

The following procedures must be followed:

Recognition/Identification of hazards   Recognition of the hazards implies some form of safety monitoring, such as a safety inspection or audit, together with feedback from accident investigation in certain cases.

Assessment and evaluation of the risks   Risk assessment requires a measurement of the magnitude of the risk based on factors such as probability or likelihood of the risk arising, the severity of outcome, in terms of injury, damage or loss, the frequency of the risk arising and the number of people exposed to the risk. Following assessment, evaluation of risk must take into account the current legislation applying to that particular risk situation.

Implementation of a control strategy   Once the risk has been assessed, it must either be eliminated or controlled. Elimination or avoidance of the risk may not be possible for a variety of reasons and, inevitably, some form of control must be implemented.

Monitoring of control strategy   It is essential that any control strategy applied is subject to regular monitoring to ensure continuing effectiveness and use of the control.

PREVENTION AND CONTROL STRATEGIES IN ACCIDENT PREVENTION

Prohibition/Elimination   This is the most extreme control strategy that can be applied, in particular where there is no known form of operator protection available, e.g. in the case of potential exposure to carcinogenic substances, or where there is an unacceptable level of risk in certain activities.

Substitution   This implies the substitution, for instance, of a less dangerous substance for a more dangerous one, or of a less dangerous system of work for a more dangerous one.

Change of process   Design or process engineering can usually change a process to afford better operator protection.

Controlled operation   This can be achieved through isolation of a particularly hazardous operation, the use of *Permit to Work Systems*, *Method Statements*, mechanical or remote control handling systems, machinery guarding, restriction of certain operations to highly trained operators, i.e. competent and/or authorised persons, and (in the case of hazardous airborne contaminants) the use of various forms of arrestment equipment.

Limitation   The limitation of exposure of personnel to specific environmental and chemical risks (e.g. noise, gases, fumes, on a time-related basis) may be appropriate in certain cases.

Ventilation   The operation of mechanical ventilation systems (e.g. receptor systems and captor systems) which remove airborne contaminants at the point of generation, or which dilute the concentration of potentially hazardous atmospheres with ample supplies of fresh air (dilution ventilation) is generally required where substances are known to be hazardous to health.

Housekeeping, personal hygiene and welfare amenity provisions   Poor levels of housekeeping are a contributory factor in many accidents. The maintenance of high standards of housekeeping is vital, particularly where flammable wastes may be produced and stored. Staff must be trained in maintaining good standards of personal hygiene, particularly where they may be handling hazardous substances. The provision of suitable and sufficient sanitary accommodation, washing and showering arrangements, facilities for clothing storage and the taking of meals must be considered.

Personal protective equipment (PPE)   The provision and use of various items of PPE (e.g. safety boots, eye protectors, gloves, etc.) is a commonly used strategy. It has severe limitations in that an operator must wear the PPE correctly all the time he is exposed to the risk. The provision and use of any item of PPE must be viewed as the last resort, when all other strategies have failed, or an interim measure until some other form of control strategy can be applied. The limitations of PPE should be clearly established and systems for maintenance and cleaning of same established and implemented.

Employers should ensure that PPE is 'suitable' in that it is appropriate for the risks and conditions where exposure may occur, takes account of ergonomic requirements and the state of health of the wearer, is capable of fitting the wearer correctly and is effective in preventing or adequately controlling the risks without increasing the overall risk.

Health surveillance   Health surveillance implies monitoring the health of identified persons on a regular basis. It may include the exclusion of certain people from specific processes or practices (e.g. women and young people) medical surveillance of certain personnel, medical examinations, health checks, health supervision, biological monitoring (e.g. blood tests, urine tests, and other forms of testing, such as audiometry).

Information, instruction and training   The provision of information to staff and the instruction and training of specific management, safety personnel and operators in the recognition of risk and the assessment of same is crucial to the success of any accident prevention programme. Staff must know why certain management action is taken and orders given, and must be fully aware that their co-operation is needed to make the workplace a safe and healthy one for themselves and others.

THE ROLE OF THE HEALTH AND SAFETY ADVISOR

Whilst the duties and accountabilities of health and safety practitioners vary from organisation to organisation, the following principal accountabilities of such persons should be considered.

1.   Advise senior and local management on measures to reduce accidents and occupational ill-health as cost effectively as possible and in accordance with legal requirements.

2.   Advise senior and local management on the health and safety management and control of contractor activities and on the acquisition and installation of new plant and equipment.

3.   Formulate and agree with directors a 'Statement of Health and Safety Policy', and monitor the implementation of policy throughout the organisation.

4.   Plan, develop and implement programmes of health and safety training, together with reviewing the feedback from such training.

5.   Develop internal *Codes of Practice* and other documentation to ensure uniformity of approach to specific issues, practices, systems and procedures throughout the organisation, monitoring the implementation of such codes of practice as necessary.

6. Ensure constructive liaison and co-operation with enforcement officers. safety organisations, insurance companies and other interested parties.

7. Promote safety, health and welfare as an integral and significant feature of the business operation.

8. Advise senior and local management on the implications and implementation of current and new health and safety legislation.

9. Provide and disseminate information on health and safety-related issues.

10. Ensure efficient investigation of all loss-producing incidents and compliance with current notification and reporting procedures.

HEALTH AND SAFETY PROFESSIONALS

Occupational health and safety is a multi-disciplinary area of study, entailing an understanding of many subject areas, such as law, toxicology, human factors, engineering, risk management, occupational hygiene and fire protection, each of which is an area of study in its own right. Health and safety professionals, therefore, must meet a high level of academic attainment in order to be in a position to advise employers on these issues. Effective occupational safety and health management requires fully trained, competent and experienced professionals capable of taking on a range of responsibilities from risk assessment to advising on health and safety policy.

The Institution of Occupational Safety and Health (IOSH) is the principal professional institution in the UK for health and safety practitioners. As such, this Institution represents over 18,000 individual members working across the full spectrum of industry and commerce – from multinationals to small consultancies. The IOSH is the focus for important matters affecting the profession. It is consulted by government departments for the views of its members on draft legislation, approved codes of practice and guidance notes, and is represented on the committees of national and international standards-making bodies. In addition, the IOSH periodically publishes policy statements on key topics and issues. Close relationships are maintained with organisations with common interests throughout the world. The Institution works to promote excellence in the discipline and practice of occupational safety and health and helps its members working in this specialist sphere of work by:

(a) granting corporate membership as a mark of technical and professional competence and the attainment of experience;

(b) providing facilities to maintain and enhance professional skills and knowledge.

Register of Safety Practitioners

The IOSH'S Register of Safety Practitioners (RSP) provides a means of identifying and measuring practical ability in addition to professional occupational safety and health knowledge. It provides employers, clients and regulatory bodies with an accepted standard of competence and capability for general safety practitioners. Entrance to the Register is only open to corporate members of the IOSH.

Qualifications and Experience

Membership of the IOSH is at both corporate and non-corporate level. Entrance at corporate level depends upon a combination of academic qualifications, experience and achievement. It is open to those holding an appropriate qualification coupled with a minimum of three years' professional experience. Appropriate qualifications are:

(a) an accredited degree or diploma in occupational safety and health or a related discipline;

(b) the National Examination Board in Occupational Safety and Health (NEBOSH) Diploma;

(c) Level 4 of the Vocational Qualifications for Occupational Health and Safety Practice.

People whose roles include some health and safety responsibilities, for example, assisting more highly qualified occupational safety and health professionals, or dealing with routine matters in low risk sectors, join the IOSH at Associate level. An Associate must hold a NEBOSH Certificate or equivalent. Affiliate level of the IOSH is designed for those who have an active interest in occupational safety and health, but who are not eligible to join at other grades of membership.

SUMMARY – CHAPTER 3

1. The term 'accident' can be defined in a number of ways.

2. Accidents are concerned with the objective danger at a point in time and the perception of risk on the part of the individual.

3. No two people perceive risk in the same way.

4. Accident prevention is directed at developing and implementing a range of *Safe Place* and *Safe Person* strategies.

5. Organisations need well-developed emergency procedures.

6. The causes of accidents are many and varied.

7. A wide range of prevention and control strategies in accident prevention are available.

8. Any competent person/health and safety specialist should have the appropriate skill, knowledge and experience to advise management on the preventive and protective measures necessary.

CHAPTER 4

HEALTH AND SAFETY MONITORING

Health and safety performance should be monitored through one or more systems, such as safety audits, safety surveys, safety inspections, safety sampling systems or safety tours. These various form of safety monitoring are described below.

Safety audits   A safety audit subjects each area of an organisation's activities to a systematic critical examination with the principal objective of minimizing loss. It is an on-going process aimed at ensuring effective health and safety management. Every component of the total system is included, e.g. management policy, attitudes, training, features of processes, training needs, emergency procedures, health surveillance arrangements, etc. Safety audits are generally carried out using a pre-designed audit checklist to ensure consistency of approach. (See Figure 9 below.)

Safety surveys   A safety survey is a detailed examination of a number of critical areas of operation (e.g. transport safety, machinery safety) or an in-depth study of the whole health and safety operation of a particular location or premises. Such an in-depth survey would consider management and administration procedures, environmental factors, occupational health and hygiene arrangements, the broad field of safety and accident prevention and arrangements for health and safety training. It is standard practice, once the survey has been completed, for management to be presented with a phased programme of health and safety improvement covering a five-year period.

Safety inspections   This is the most commonly used form of safety monitoring and is a scheduled inspection of premises or part of same by various persons, e.g. managers, safety specialists, trade union safety representatives or members of a safety committee.

Safety tours   These are an unscheduled examinations of a work area, carried out by a manager, possibly accompanied by safety committee members, to ensure, e.g. that standards of housekeeping are at an acceptable level, fire protection measures are being observed or personal protective equipment is being used correctly. A safety tour tends to be spontaneous as opposed to, say an audit or survey, which may be planned some time ahead.

Damage control   Damage control and costing techniques emphasise the fact that non-injury accidents are as important as injury accidents. The elimination of non-injury accidents will, in many cases, remove the potential for other forms of accident which result in injury. Evidence of damage to property, machinery and plant is frequently an indication of poor safety performance. The operation of a *Hazard Reporting System* is essential.

Safety sampling   This technique is designed to measure, by random sampling, the accident potential in a specific workplace or at a particular process by identifying safety defects or omissions. Safety sampling entails the use of a safety sampling sheet with a limited number of aspects to be observed, e.g. machinery guards in position, ear protection being worn, housekeeping being maintained. (See Figure 10 below.) Each of the items for consideration is graded according to significance and a maximum number of points are awardable for each aspect. Such a system monitors the effectiveness of the overall programme.

Performance Monitoring and Review

Monitoring implies the continuing assessment of the performance of the organisation, of the tasks that people undertake, and of managers and operators on an individual basis, against agreed objectives and standards. Performance monitoring is a common feature of most organisations in terms of financial performance, production and sales performance. It implies clear identification of the organisation's objectives, perhaps through a written mission statement and the setting of polices and objectives which are both measurable and achievable by all parts of the organisation, supported by adequate resources. Performance monitoring is generally accepted as a standard feature of good management practice. Current legislation clearly emphasises the need to monitor performance. This may take place through the setting of health and safety policy, organisational development or re-development, risk assessment, the establishment of the role and functions of competent persons, or following the actual development of techniques of planning, measuring and reviewing performance.

At an organisational level, senior management should be aware of the various strengths and weaknesses in health and safety performance. This may be identified through various forms of active safety monitoring and through reactive monitoring systems, such as accident investigation and the analysis of accident and sickness absence returns. At a task level, the implementation of formally-established safe systems of work, permit to work systems, in-company codes of practice and operating instructions are important indicators of performance. Reactive monitoring through feedback from training exercises, in particular those which are aimed at increasing people's awareness, in improving attitudes to safe working and generally raising levels of knowledge of hazards, will indicate whether there has been an improvement in performance or not.

Safety Performance and the Reward Structure of the Organisation

Safety performance should be related to the reward structure of the organisation in order to produce the right motivation amongst all levels of management. Most organisations practising *management by objectives* or *performance-related pay* have a standard form of job and career review or annual appraisal for staff, but how many of these systems take into account the health and safety performance of the various levels of management?

On-the-job performance monitoring should take into account the human decision-making components of a job, in particular the potential for human error. Is there a need for job safety analysis leading to the formulation of job safety instructions, a review of current job design and/or examination of the environmental factors surrounding the job?

And what about the operator? Current legislation places an absolute duty on employers to take into account the capabilities of employees when entrusting them with tasks. The whole concept of individual capability for the safe operation of a task is extremely broad. Here there is a need to consider not only physical capability to, for instance, load and unload products in and out of vehicles, but also mental capability in terms of the degree of understanding necessary for certain tasks. The latter may be achieved through the various stages of training aimed at ensuring competence for tasks and in continuing assessment of operators by line management.

Figure 9: Safety Audit

Documentation	Yes/No
1. Are management aware of all health and safety legislation applying to their workplace?	
Is this legislation available to management and employees?	
2. Have all approved codes of practice, HSE Guidance Notes and internal codes of practice been studied by management with a view of ensuring compliance?	
3. Does the existing policy statement meet current conditions in the workplace?	
Is there a named director/senior manager with overall responsibility for health and safety?	
Are the organisation and arrangements to implement the safety policy still adequate?	

	Yes/No

Have the hazards and precautions necessary on the part of staff and other persons been identified and recorded?

Are individual responsibilities for health and safety clearly detailed in the policy statement?

4. Do all job descriptions adequately describe individual health and safety responsibilities and accountabilities?

5. Do written safe systems of work exist for all potentially hazardous operations?

Is 'permit to work' documentation available?

6. Has a suitable and sufficient assessment of the risks to staff and other persons been made, recorded and brought to the attention of staff and other persons?

Have other risk assessments in respect of:

(a) substances hazardous to health;
(b) risks to hearing;
(c) work equipment;
(d) personal protective equipment;
(e) manual handling operations;
(f) display screen equipment;

been made, recorded and brought to the attention of staff and other persons?

7. Is there a record of inspections of the means of escape in the event of fire, fire appliances, fire alarms, warning notices, fire and smoke detection equipment?

8. Is there a record of inspections and maintenance of work equipment, including guards and safety devices?

Are all examination and test certificates available, e.g. lifting appliances and pressure systems?

9. Are all necessary licences available, e.g. to store petroleum spirit?

10. Are workplace health and safety rules and procedures available, promoted and enforced?

Have these rules and procedures been documented in a way which is comprehensible to staff and others, e.g. a 'Health and Safety Handbook'?

Are disciplinary procedures for unsafe behaviour clearly documented and known to staff and other persons?

11. Is a formally written emergency procedure available?

12. Is documentation available for the recording of injuries, near misses, damage only accidents, diseases and dangerous and dangerous occurrences?

13. Are health and safety training records maintained?

14. Are there documented procedures for regulating the activities of contractors, visitors and other persons working on the site?

15. Is hazard reporting documentation available to staff and other persons?

16. Is there a documented planned maintenance system to cover the workplace and work equipment?

17. Are there written cleaning schedules?

Health and Safety Systems

1. Have competent persons been appointed to:

   (a) co-ordinate health and safety measures?
   (b) implement the emergency procedure?

   Have these persons been adequately trained on the basis of identified and assessed risks?

   Are the role, function, responsibility and accountability of competent persons clearly identified?

2. Are there arrangements for specific forms of safety monitoring, e.g. safety inspections, safety sampling?

   Is a system in operation for measuring and monitoring individual management performance on health and safety issues?

3. Are systems established for the formal investigation of accidents, ill-health, near misses and dangerous occurrences?

   Do investigation procedures produce results which can be used to prevent future incidents?

   Are the causes of accidents, ill-health, near misses and dangerous occurrences analysed in terms of failure of established safe systems of work?

	Yes/No

4. Is a hazard reporting system in operation?

5. Is a system for controlling damage to structural items, machinery, vehicles, etc. in operation?

6. Is the system for joint consultation with a safety representatives and staff effective?

   Are the role, constitution and objectives of the Health and Safety Committee clearly identified?

   Are the procedures for appointing or electing committee members and trade union safety representatives clearly identified?

   Are the available facilities, including training arrangements, known to committee members and trade union safety representatives?

7. Are the capabilities of employees as regards health and safety taken into account when entrusting them with tasks?

8. Is the provision of first aid arrangements adequate?

9. Are the procedures covering sickness absence known to staff?

   Is there a procedure for controlling sickness absence?

   Are managers aware of the current sickness absence rate?

10. Do current arrangements ensure that health and safety implications are considered at the design stage of projects?

11. Is there a formally-established annual health and safety budget?

    Are first aid personnel adequately trained and re-trained?

Prevention and Control Procedures

1. Are formal inspections of machinery, plant, hand tools, access equipment, electrical equipment, storage equipment, warning systems, first aid boxes, resuscitation equipment, welfare amenity areas, etc. undertaken?

   Are machinery guards and safety devices examined on a regular basis?

2. Is a 'permit to work' system operated where there is a high degree of foreseeable risk?

3. Are fire and emergency procedures practised on a regular basis?

   Where specific fire hazards have been identified, are they catered for in the current fire protection arrangements?

Are all items of fire protection equipment and alarms tested, examined and maintained on a regular basis?

Are all fire exits and escape routes marked, kept free from obstruction and operational?

Are all fire appliances correctly labelled, sited and maintained?

4. Is a planned maintenance system in operation?

5. Are the requirements of cleaning schedules monitored?

Is housekeeping of a high standard, e.g. material storage, waste disposal, removal of spillages?

Are all gangways, stairways, fire exits, access and egress points to the workplace maintained and kept clear?

6. Is environmental monitoring of temperature, lighting, ventilation, humidity, radiation, noise and vibration undertaken on a regular basis?

7. Is health surveillance of persons exposed to assessed health risks undertaken on a regular basis?

8. Is monitoring of personal exposure to assessed health risks undertaken on a regular basis?

9. Are local exhaust ventilation systems examined, tested and maintained on a regular basis?

10. Are arrangements for the storage and handling of substances hazardous to health adequate?

Are all substances hazardous to health identified and correctly labelled, including transfer containers?

11. Is the appropriate personal protective equipment available?

Is the personal protective equipment worn or used by staff consistently when exposed to risks?

Are storage facilities for items of personal protective equipment provided?

12. Are welfare amenity provisions, i.e. sanitation, hand washing, showers and clothing storage arrangements, adequate?

Do welfare amenity provisions promote appropriate levels of personal hygiene?

	Yes/No

Information, Instruction, Training and Supervision

1. Is the information provided by manufacturers and suppliers of articles and substances for use at work adequate?

    Do employees and other persons have access to this information?

2. Is the means of promoting health and safety adequate?

    Is effective use made of safety propaganda, e.g. posters?

3. Do current safety signs meet the requirements of current legislation?

    Are safety signs adequate in terms of the assessed risks?

4. Are fire instructions prominently displayed?

5. Are hazard warning systems adequate?

6. Are the individual training needs of staff and other persons assessed on a regular basis?

7. Is staff health and safety training undertaken:

    (a) at the induction stage;

    (b) on their being exposed to new or increased risks because of:

    (i)   transfer or change in responsibilities;

    (ii) the introduction of new work equipment or a change respecting existing work equipment;

    (iii) the introduction of new technology;

    (iv) the introduction of a new system of work or change in an existing system of work.

    Is the above training:

    (a) repeated periodically?

    (b) adapted to take account of new or changed risks?

    (c) carried out during working hours?

8. Is specific training carried out regularly for first aid staff, fork lift truck drivers, crane drivers and others exposed to specific risks?

    Are selected staff trained in the correct use of fire appliances?

	Yes/No

Final Question

Are you satisfied that your organisation is as safe and healthy as you can reasonably make it, or that you know what action must be taken to achieve that state?

Action Plan

1. Immediate action

2. Short-Term Action (14 days)

3. Medium-Term Action (6 months)

4. Long-Term Action (2 years)

Auditor _____   Date _____

NOTIFICATION AND REPORTING OF INJURIES, DISEASES AND DANGEROUS
OCCURRENCES

The Reporting of Injuries, Diseases and Dangerous Occurrences regulations 1995 (RIDDOR) cover the requirement to notify and report certain categories of injury and disease sustained by people at work, together with specified dangerous occurrences and gas incidents to the relevant enforcing authority – i.e. HSE or Local Authority. The majority of duties in 'responsible persons' (as defined) are of an absolute nature.

The Principal Requirements of RIDDOR

1. Regulation 3 The responsible person (e.g. employer) is to notify the relevant enforcing authority by the quickest practicable means and subsequently make a report within ten days on the approved form in respect of death, and defined major injury and dangerous occurrence arising out of, or in connection with, work.

2. Regulation 4 The duty of an employer to report the death of an employee where, as a result of an accident at work, the injured employee dies within one year of the accident.

Figure 10: Safety Sampling Exercise

		Area A	Area B	Area C	Area D	Area E
1. Housekeeping/cleaning	(Max 10)					
2. Personal protection	(Max 10)					
3. Machinery safety	(Max 10)					
4. Chemical storage	(Max 5)					
5. Chemical handling	(Max 5)					
6. Manual handling	(Max 5)					
7. Fire protection	(Max 10)					
8. Structural safety	(Max 10)					
9. Internal transport	(Max 5)					
10. Access equipment	(Max 5)					
11. First aid provision	(Max 5)					
12. Hand tools	(Max 10)					
13. Internal storage	(Max 5)					
14. Electrical safety	(Max 10)					
15. Temperature control	(Max 5)					
16 Lighting	(Max 5)					
17. Ventilation	(Max 5)					
18. Noise control	(Max 5)					
19. Dust/fume control	(Max 5)					
20. Welfare amenities	(Max 10)					
TOTAL	(Max 140)					

3. Regulation 5   The duty of an employer to report cases of any disease (listed in the Schedule) to employees.

4. Regulation 6   The duties on certain persons to report gas incidents.

5. Regulation 7   Responsible persons to keep records of reportable injuries, diseases and dangerous occurrences.

Figure 11: Report of an injury or dangerous occurrence

Health and Safety at Work etc Act 1974
The Reporting of Injuries, Diseases and Dangerous Occurrences Regulations 1995

HSE
Health & Safety
Executive

Report of an injury or dangerous occurrence

Filling in this form
This form must be filled in by an employer or other responsible person.

Part A

About you

1 What is your full name?

2 What is your job title?

3 What is your telephone number?

About your organisation

4 What is the name of your organisation?

5 What is its address and postcode?

6 What type of work does the organisation do?

Part B

About the incident

1 On what date did the incident happen?

    /   /

2 At what time did the incident happen?
(Please use the 24-hour clock eg 0600)

3 Did the incident happen at the above address?

Yes ☐ Go to question 4

No ☐ Where did the incident happen?
- ☐ elsewhere in your organisation – give the name, address and postcode
- ☐ at someone else's premises – give the name, address and postcode
- ☐ in a public place – give details of where it happened

If you do not know the postcode, what is the name of the local authority?

4 In which department, or where on the premises, did the incident happen?

Part C

About the injured person

If you are reporting a dangerous occurrence, go to Part F.
If more than one person was injured in the same incident, please attach the details asked for in Part C and Part D for each injured person.

1 What is their full name?

2 What is their home address and postcode?

3 What is their home phone number?

4 How old are they?

5 Are they
- ☐ male?
- ☐ female?

6 What is their job title?

7 Was the injured person (tick only one box)
- ☐ one of your employees?
- ☐ on a training scheme? Give details:

- ☐ on work experience?
- ☐ employed by someone else? Give details of the employer:

- ☐ self-employed and at work?
- ☐ a member of the public?

Part D

About the injury

1 What was the injury? (eg fracture, laceration)

2 What part of the body was injured?

F2508 (01/96)   Continued overleaf

Figure 11: Report of an injury or dangerous occurrence (continued)

3 Was the injury (tick the one box that applies)
- [ ] a fatality?
- [ ] a major injury or condition? (see accompanying notes)
- [ ] an injury to an employee or self-employed person which prevented them doing their normal work for more than 3 days?
- [ ] an injury to a member of the public which meant they had to be taken from the scene of the accident to a hospital for treatment?

4 Did the injured person (tick all the boxes that apply)
- [ ] become unconscious?
- [ ] need resuscitation?
- [ ] remain in hospital for more than 24 hours?
- [ ] none of the above.

Part E

About the kind of accident
Please tick the one box that best describes what happened, then go to Part G.

- [ ] Contact with moving machinery or material being machined
- [ ] Hit by a moving, flying or falling object
- [ ] Hit by a moving vehicle
- [ ] Hit something fixed or stationary

- [ ] Injured while handling, lifting or carrying
- [ ] Slipped, tripped or fell on the same level
- [ ] Fell from a height
  - How high was the fall?
  - [_____] metres

- [ ] Trapped by something collapsing

- [ ] Drowned or asphyxiated
- [ ] Exposed to, or in contact with, a harmful substance
- [ ] Exposed to fire
- [ ] Exposed to an explosion

- [ ] Contact with electricity or an electrical discharge
- [ ] Injured by an animal
- [ ] Physically assaulted by a person

- [ ] Another kind of accident (describe it in Part G)

Part F

Dangerous occurrences
Enter the number of the dangerous occurrence you are reporting. (The numbers are given in the Regulations and in the notes which accompany this form.)

Part G

Describing what happened
Give as much detail as you can. For instance
- the name of any substance involved
- the name and type of any machine involved
- the events that led to the incident
- the part played by any people.

If it was a personal injury, give details of what the person was doing. Describe any action that has since been taken to prevent a similar incident. Use a separate piece of paper if you need to.

Part H

Your signature
Signature

Date
[/ /]

Where to send the form
Please send it to the Enforcing Authority for the place where it happened. If you do not know the Enforcing Authority, send it to the nearest HSE office.

For official use
Client number | Location number | Event number
[ ] INV REP [ ] Y [ ] N

Figure 12: Report of a case of disease

Health and Safety at Work etc Act 1974
The Reporting of Injuries, Diseases and Dangerous Occurrences Regulations 1995

HSE
Health & Safety Executive

Report of a case of disease

Filling in this form
This form must be filled in by an employer or other responsible person.

Part A

About you

1 What is your full name?

2 What is your job title?

3 What is your telephone number?

About your organisation

4 What is the name of your organisation?

5 What is its address and postcode?

6 Does the affected person usually work at this address?
Yes ☐ Go to question 7
No ☐ Where do they normally work?

7 What type of work does the organisation do?

Part B

About the affected person

1 What is their full name?

2 What is their date of birth?
/ /

3 What is their job title?

4 Are they
☐ male?
☐ female?

5 Is the affected person (tick one box)
☐ one of your employees?
☐ on a training scheme? Give details:

☐ on work experience?
☐ employed by someone else? Give details:

☐ other? Give details:

F2508A (01/96)

Continued overleaf

Figure 12: Report of a case of disease (continued)

Part C

The disease you are reporting

1 Please give:

• the name of the disease, and the type of work it is associated with; or

• the name and number of the disease (from Schedule 3 of the Regulations – see the accompanying notes).

Continue your description here

2 What is the date of the statement of the doctor who first diagnosed or confirmed the disease?

/ /

3 What is the name and address of the doctor?

Part D

Describing the work that led to the disease

Please describe any work done by the affected person which might have led to them getting the disease.

If the disease is thought to have been caused by exposure to an agent at work (eg a specific chemical) please say what that agent is.

Give any other information which is relevant.

Give your description here

Part E

Your signature

Signature

Date

/ /

Where to send the form

Please send it to the Enforcing Authority for the place where the affected person works. If you do not know the Enforcing Authority, send it to the nearest HSE office.

For official use

Client number

Location number

Event number

☐ INV REP ☐ Y ☐ N

Notifiable and Reportable Major Injuries

These are listed in Schedule 1 of the RIDDOR as follows:

1. Any fracture, other than to the fingers, thumbs or toes.

2. Any amputation.

3. Dislocation of the shoulder, hip or knee.

4. Loss of sight (whether temporary of permanent).

5. A chemical or hot metal burn to the eye or any penetrating injury to the eye.

6. Any injury resulting from electric shock or electrical burn (including any electrical burn caused by arcing or arcing products) leading to unconsciousness or requiring resuscitation or admittance to hospital for more than 24 hours.

7. Any other injury:

   (a) leading to hypothermia, heat induced illness or to unconsciousness;

   (b) requiring resuscitation;

   (c) or requiring admittance to hospital for more than 24 hours.

8. Loss of consciousness caused by asphyxia or by exposure to an harmful substance or biological agent.

9. Either of the following conditions which result from the absorption of any substance by inhalation, ingestion or through the skin:

   (a) acute illness requiring medical treatment;

   (b) loss of consciousness.

10. Acute illness which requires medical treatment where there is reason to believe that this resulted from exposure to a biological agent or its toxins or infected material.

Scheduled Dangerous Occurrences

A dangerous occurrence is a major incident as a rule that has the potential for significant damage and potential loss of life and which is listed in Schedule 2 of the RIDDOR. Dangerous occurrences are classified under five headings.

1. **General** e.g. incidents involving lifting machinery, pressure systems, overhead electric lines.

2. **Dangerous occurrences in mines that are reportable** e.g. fire or ignition of gas, escape of gas, insecure tip.

3. **Dangerous occurrences in respect of quarries that are reportable** e.g. misfires, movement of slopes or faces.

4. **Dangerous occurrences in respect of relevant transport systems that are reportable** e.g. accidents involving any kind of train, incidents at level crossings.

5. **Dangerous occurrences in respect to an offshore workplace that are reportable** e.g. releases of petroleum hydrocarbon, fire or explosion.

Reportable Diseases

These are diseases listed in Schedule 3 of the RIDDOR under three classifications.

1. **Conditions due to physical agents and the physical demands of work** e.g. malignant diseases of bones due to ionising radiation, decompression illness.

2. **Infections due to biological agents** e.g. anthrax, brucellosis, leptospirosis.

3. **Conditions due to substances** e.g. poisoning by carbon dispulphide, ethylene oxide and methyl bromide.

THE INVESTIGATION OF ACCIDENTS

Investigation of the direct and indirect causes of accidents is a reactive strategy in safety management. There are very good reasons, however, for the effective and thorough investigation of accidents, viz:

(a) on a purely humanitarian basis, no one likes to see people killed or injured;

(b) the accident may have resulted from a breach of statute or regulations by the organisation, the accident victim, the manufacturers and/or suppliers of articles and substances used at work, or other persons, e.g. contractors, with the possibility of civil proceedings being instituted by the injured party against the employer and other persons;

(c) the injury may be reportable to the enforcing authority (HSE, Local Authority) under the current legislation;

(d) the accident may result in lost production;

(e) from a management viewpoint, a serious accident, particularly a fatal one, can have a long-term detrimental effect on the morale of the work force and management/worker relations;

(f)	there may be damage to plant and equipment, resulting in the need for repair or replacement, with possible delays in replacement;

(g)	in most cases, there will be a need for immediate remedial action in order to prevent a recurrence of the accident.

Apart from the obvious direct and indirect losses associated with all forms of incident, not just those resulting in injury, there are legal reasons for investigating accidents to identify the direct and indirect causes and to produce strategies for preventing recurrences. Above all, the purpose of accident investigation is not to apportion blame or fault, although blame or fault may eventually emerge as a result of accident investigation.

What Accidents should be Investigated?

Clearly there is a case for investigating all accidents and, indeed, 'near misses'. A near miss is defined as "an unplanned and unforeseeable event that could have resulted, but did not result, in human injury, property damage or other form of loss". However, it may be impracticable to investigate every accident, but the following factors should be considered in deciding which accidents should be investigated as a priority:

(a)	the type of accident, e.g. fall from a height, chemical handling, machinery-related;

(b)	the form and severity of injury, or the potential for serious injury and/or damage;

(c)	whether the accident indicates the continuation of a particular trend in the organisation's accident experience;

(d)	the extent of involvement of articles and substances used at work, e.g. machinery, work equipment, hazardous substances, and the ensuing damage or loss;

(e)	the possibility of a breach of the law;

(f)	whether the injury or occurrence is, by law, notifiable and reportable to the enforcing authority;

(g)	whether the accident should be reported to the insurance company as it could result in a claim being submitted.

Practical Accident Investigation

In any serious incident situation, such as a fatal accident, an accident resulting in major injury (e.g. fractures, amputations, loss of an eye), or where there has been a scheduled dangerous occurrence, such as the collapse of a crane, speed of action is essential. This is particularly the case when it comes to interviewing injured persons and witnesses.

The following procedure is recommended:

(a) establish the facts as quickly and completely as possible with regard to:
  (i) the general environment;
  (ii) the particular plant, machinery, practice or system of work involved;
  (iii) the sequence of events leading to the accident;

(b) use an instant camera to take photographs of the accident scene prior to any clearing up that may follow the accident;

(c) draw sketches and take measurements with a view to producing a scale drawing of the events leading up to the accident;

(d) list the names of all witnesses, i.e. those who saw, heard, felt or smelt anything; interview them thoroughly in the presence of a third party if necessary, and take full statements. In certain cases, it may be necessary to formally caution witnesses prior to their making a statement. Do not prompt or lead witnesses;

(e) evaluate the facts, and individual witnesses' versions of them, as to accuracy, reliability and relevance;

(f) endeavour to arrive at conclusions as to the causes of the accident on the basis of the relevant facts;

(g) examine closely any contradictory evidence. Never dismiss a fact that does not fit in with the rest – find out more;

(h) learn fully about the system of work involved. Every accident occurs within the context of a work system. Consider the people involved in terms of their ages, training, experience and level of supervision, and the nature of the work, e.g. routine, sporadic or incidental;

(i) in certain cases, it may be necessary for plant and equipment, such as lifting appliances, machinery and vehicles to be examined by a specialist, e.g. a consultant engineer;

(j) produce a report for the responsible manager emphasising the causes and remedies to prevent a recurrence, including any changes necessary;

(k) in complex and serious cases, consider the establishment of an investigating committee comprising managers, supervisors, technical specialists and trade union safety representatives.

There are other persons who may also need to investigate this accident, including:

(a) trade union safety representatives, should a member of their trade union be injured;

(b) insurance company liability surveyor, in the event of a claim being submitted by the organisation;

(c) legal representative of the injured party to establish the cause of the accident and whether there has been a breach of statutory duty or negligence on the part of the employer;

(d) officers of the enforcing authority, to establish whether there has been a breach of current health and safety legislation which may require action, such as the service of a Prohibition or Improvement Notice or prosecution. It is essential, therefore, that any accident report produced is accurate, comprehensible and identifies the causes of the accident, both direct and indirect.

The Outcome of Accident Investigation

Whether the investigation of an accident is undertaken by an individual, e.g. health and safety specialist, or by a special committee, it is necessary, once the causes have been identified, to submit recommendations to management with a view to preventing a recurrence. The organisation of feedback on the causes of accidents is crucial in large organisations, especially those who operate more than one site or premises.

An effective investigation should result in one or more of the following recommendations being made:

(a) the issuing of specific instructions by management covering, for instance, systems of work, the need for more effective guarding of machinery or safe manual handling procedures;

(b) the establishment of a working party or committee to undertake further investigation, perhaps in conjunction with members of the safety committee and/or safety representatives;

(c) the preparation and issue of specific codes of practice or guidance notes dealing with the procedures necessary to minimize a particular risk, e.g. the use of a 'permit to work' system;

(d) the identification of specific training needs for groups of individuals (e.g. managers, foremen, supervisors, machinery operators, drivers) and the implementation of a training programme designed to meet these needs;

(e) the formal analysis of the job or system in question, perhaps using job safety analysis techniques, to identify skill, knowledge and safety components of the job;

(f) identification of the need for further information relating to articles and substances used at work, e.g. work equipment, chemical substances;

(g) identification of the need for better environmental control, e.g. noise reduction at source or improved lighting;

(h) general employee involvement in health and safety issues, e.g. the establishment of a health and safety committee;

(j) identification of the specific responsibilities of groups with regard to safe working practices.

Above all, a system of monitoring should be implemented to ensure that the lessons which have been learned from the accident are put into practice or incorporated in future systems of work, and that procedures and operating systems have been produced for all grades of staff.

STATISTICAL INFORMATION

The use of statistics is an important feature of health and safety management, provided the limitations of same are recognised. In particular, due to under-reporting of accidents and occupational ill-health, they may not be a true indicator of health and safety performance. Their principal use is in the indication of trends in accident and ill-health experience.

Accident and Ill-Health Rates

The following rates are used in the calculation of accident and ill-health rates:

$$\text{Frequency Rate} = \frac{\text{Total number of accidents}}{\text{Total number of man-hours worked}} \times 100{,}000$$

$$\text{Incident Rate} = \frac{\text{Total number of accidents}}{\text{Number of persons employed}} \times 100{,}000$$

$$\text{Severity Rate} = \frac{\text{Total number of days lost}}{\text{Total number of man-hours worked}} \times 100{,}000$$

$$\text{Mean Duration Rate} = \frac{\text{Total number of days lost}}{\text{Total number of accidents}}$$

$$\text{Duration Rate} = \frac{\text{Number of man-hours worked}}{\text{Total number of accidents}}$$

$$\text{Sickness Absence Rate} = \frac{\text{Total man-days lost through sickness}}{\text{Total man-days worked}} \times 100{,}000$$

SUMMARY – CHAPTER 4

1. Organisations should operate regular health and safety monitoring procedures, such as audits, inspections and sampling exercises.

2. Management performance in health and safety should be regularly monitored and incorporated in the reward structure of the organisation.

3. Safety audits are an important means of measuring health and safety performance.

4. Formal systems for the notification and reporting of injuries, diseases and dangerous occurrences to the enforcing authority are required.

5. All accidents should be investigated with a view to identifying both direct and indirect causes.

6. The maintenance of statistical information on accidents and ill-health is an important means of monitoring safety performance.

PART 2

OCCUPATIONAL HEALTH

CHAPTER 5

OCCUPATIONAL DISEASES AND CONDITIONS

OCCUPATIONAL HEALTH AND HYGIENE – GENERAL PRINCIPLES

Every year people at work contract various forms of occupational disease or condition. Many people die as a result of contracting, for instance, occupational cancer, pneumoconiosis or chemical poisoning. Other people may be permanently incapacitated through conditions such as noise-induced hearing loss (occupational deafness), vibration-induced injury or occupational asthma. Occupational dermatitis is the most common form of occupational disease.

In the study of occupational health and hygiene it is essential to understand the meaning of these two terms.

Occupational health   This is a preventive form of medicine which examines, firstly, the relationship of work to health and, secondly, the effects of work on the worker.

Occupational hygiene   This discipline is concerned with identification, measurement, evaluation and control of contaminants and other phenomena, such as noise and radiation, which would have otherwise unacceptable adverse effects on the health of people exposed to them. Occupational hygiene practice is principally concerned with the identification, measurement, evaluation, prevention or control of a wide range of environmental stressors. These stressors can include, for instance, bacteria, gases, fumes, fogs, mists and dusts, which can affect the health of people at work. Occupational hygiene practice is also concerned with the measurement, evaluation and control of various physical phenomena, such as noise and vibration.

Occupational hygiene practice takes a number of clearly identified stages.

1.   Identification of the stressor, such as noise, dust, gases.

2. Measurement of the extent of the stressor using prescribed sampling techniques.

3. Evaluation of the risks by reference to established criteria.

4. Selection of an prevention or control strategy, such as the installation of a local exhaust ventilation system to remove dust from a process.

5. Implementation of that strategy.

6. Monitoring of the working environment to ensure the prevention or control strategy installed is effective.

THE CAUSES OF OCCUPATIONAL ILL-HEALTH

Occupational health hazards can be classified as follows:

Physical

The effects of exposure to extremes of temperature and humidity, inadequate lighting and ventilation, noise and vibration, dusts, pressure and radiation can result in a range of conditions, such as heat stroke, heat cataract, noise-induced hearing loss, vibration-induced white finger, pneumoconiosis, decompression sickness and radiation sickness.

Chemical

Exposure to toxic, corrosive, harmful or irritant solids, liquids, fumes, mists and gases is responsible for various forms of metallic and chemical poisoning, dermatitis and occupational cancers. Clearly, the form taken by a substance is significant in its potential for harm.

Biological

A range of diseases, such as various forms of human anthrax, leptospirosis, brucellosis, viral hepatitis, legionnaires' disease and aspergillosis (farmer's lung) are caused through exposure to a range of micro-organisms, such as viruses and bacteria. Certain diseases, such as brucellosis and anthrax, are transmissible from animals to people (zoonoses).

Ergonomic (work related)

The effects on people of poorly-designed working layouts and operator workstations, together with excessive and repetitive movements of joints, can lead to visual and postural fatigue, physical and mental stress and a range of conditions. These include, for instance, writer's cramp, various 'beat disorders' (beat knee, beat wrist and beat elbow) and that group of conditions known as the 'work-related upper limb disorders' or repetitive strain injury. This last-mentioned group includes conditions such as tenosynovitis and carpal tunnel syndrome. The potential for stress-induced injury must also be considered.

Various conditions arising from manual handling operations, e.g. prolapsed intervertebral discs, hernia and ligamental strain also come into this category.

OCCUPATIONAL HEALTH AND HYGIENE – FACTORS FOR CONSIDERATION

Any strategy designed to prevent or reduce the risk of people contracting various forms of ill-health at work should take the following factors into account.

1. Measures to prevent or reduce the risk of occupational disease, including various health surveillance procedures.

2. Systems for the identification, measurement, evaluation and control of occupational health risks in the working environment.

3. Welfare amenity provisions including sanitation arrangements, hand washing and shower facilities, clothing storage, taking of meals and the provision of pure drinking water.

4. First aid arrangements, including the training of first aid staff, and the provision of emergency services in the event of serious injury.

5. The ergonomic aspects of jobs.

6. The selection, provision, assessment of suitability, maintenance and use of personal protective equipment.

7. The provision of information, instruction, training and constant supervision for all persons exposed to health risks.

Practitioners in Occupational Health and Hygiene

Various specialists are involved in occupational health and hygiene practice. These include occupational physicians, occupational hygienists, occupational health nurses and health and safety practitioners. Each of these groups have a specific contribution to make in preventing ill-health arising from workplace operations and activities.

FIRST AID

First aid is defined by the RoSPA as "the skilled application of accepted principles of treatment on the occurrence of an accident or in the case of sudden illness, using facilities and materials available at the time".
   First aid is rendered:

(a) to sustain life;

(b) to prevent deterioration in an existing condition;

(c) to promote recovery.

The significant areas of first aid treatment are:

(a) restoration of breathing (resuscitation);

(b) control of bleeding;

(c) prevention of collapse.

Specific provisions relating to first aid arrangements in places of work are dealt with in the Health and Safety (First Aid) Regulations 1981 and ACOP. These include the certification and provision of occupational first aiders, the contents of first aid boxes and kits, the provision of first aid rooms. There is a general duty on employers to provide or ensure the provision of first aid at every place of work.

HEALTH SURVEILLANCE

Health surveillance is the regular review of the health of persons exposed to hazardous substances or working in specified processes. It should be carried out by a suitably qualified person, such as an occupational physician or an occupational health nurse.
   It may involve the assessment of hazardous substances or their by-products in the body by the examination of urine or blood or the assessment of body functions, such as blood pressure or lung function. In some cases, clinical tests or examinations may be necessary. Where medical examinations and inspections are required by law, employers must provide suitable facilities on site.
   Where health surveillance is required, records must be kept listing the personal details of the employee, e.g. name, sex, age and previous and current occupations which involved exposure to hazardous substances. The records, or copies of same, should be kept for at least 30 years after the date of the last entry.

Figure 13: Resuscitation Procedure Display Card
(Source Royal Society for the Prevention of Accidents)

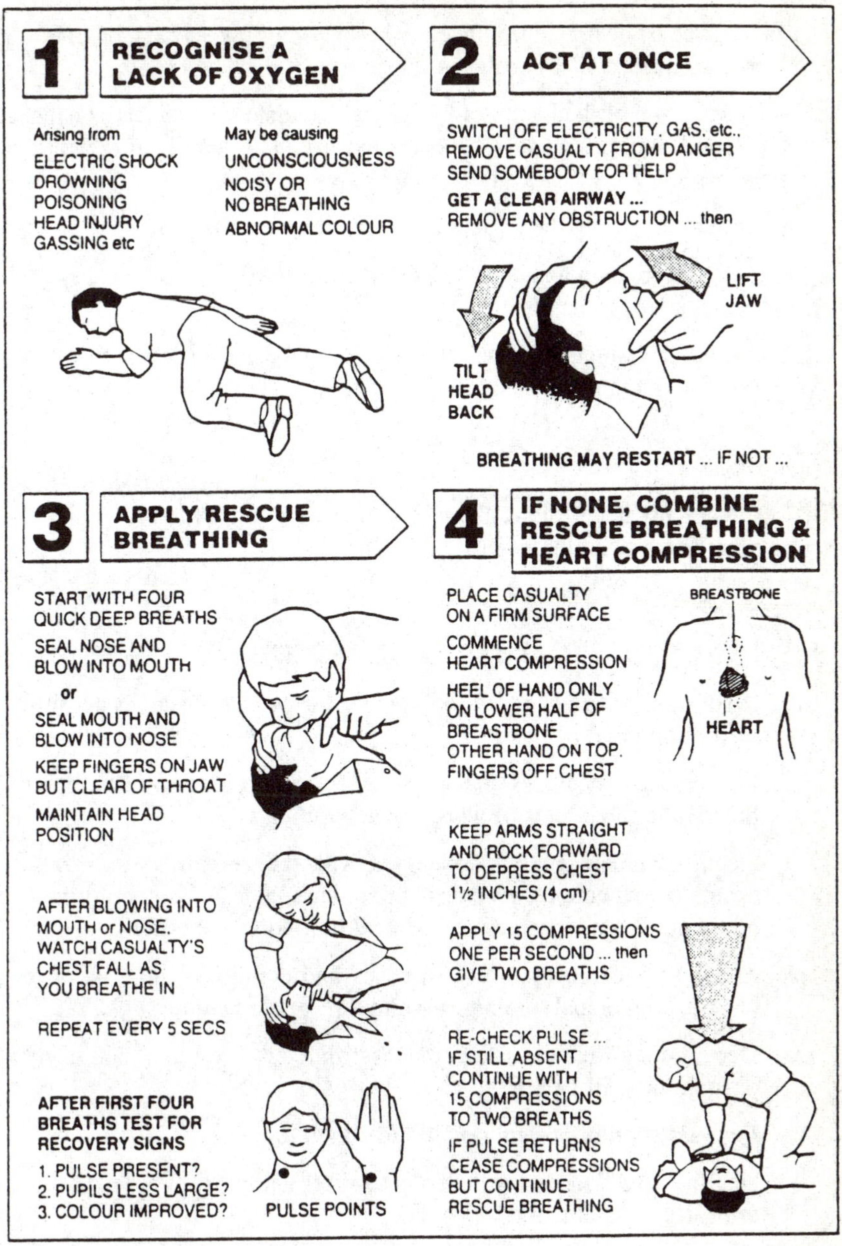

1 RECOGNISE A LACK OF OXYGEN

Arising from
ELECTRIC SHOCK
DROWNING
POISONING
HEAD INJURY
GASSING etc

May be causing
UNCONSCIOUSNESS
NOISY OR
NO BREATHING
ABNORMAL COLOUR

2 ACT AT ONCE

SWITCH OFF ELECTRICITY. GAS. etc..
REMOVE CASUALTY FROM DANGER
SEND SOMEBODY FOR HELP

GET A CLEAR AIRWAY ...
REMOVE ANY OBSTRUCTION ... then

LIFT JAW

TILT HEAD BACK

BREATHING MAY RESTART ... IF NOT ...

3 APPLY RESCUE BREATHING

START WITH FOUR
QUICK DEEP BREATHS

SEAL NOSE AND
BLOW INTO MOUTH
   or
SEAL MOUTH AND
BLOW INTO NOSE

KEEP FINGERS ON JAW
BUT CLEAR OF THROAT

MAINTAIN HEAD
POSITION

AFTER BLOWING INTO
MOUTH or NOSE.
WATCH CASUALTY'S
CHEST FALL AS
YOU BREATHE IN

REPEAT EVERY 5 SECS

**AFTER FIRST FOUR
BREATHS TEST FOR
RECOVERY SIGNS**

1. PULSE PRESENT?
2. PUPILS LESS LARGE?
3. COLOUR IMPROVED?   PULSE POINTS

4 IF NONE, COMBINE RESCUE BREATHING & HEART COMPRESSION

PLACE CASUALTY
ON A FIRM SURFACE

COMMENCE
HEART COMPRESSION

HEEL OF HAND ONLY
ON LOWER HALF OF
BREASTBONE
OTHER HAND ON TOP.
FINGERS OFF CHEST

BREASTBONE

HEART

KEEP ARMS STRAIGHT
AND ROCK FORWARD
TO DEPRESS CHEST
1½ INCHES (4 cm)

APPLY 15 COMPRESSIONS
ONE PER SECOND ... then
GIVE TWO BREATHS

RE-CHECK PULSE ...
IF STILL ABSENT
CONTINUE WITH
15 COMPRESSIONS
TO TWO BREATHS

IF PULSE RETURNS
CEASE COMPRESSIONS
BUT CONTINUE
RESCUE BREATHING

The Purpose of Health Surveillance

Objectives

The objectives of health surveillance, wherever employees may be exposed to substances hazardous to health in the course of their work, are:

(a) the protection of the health of individuals by detecting adverse changes, attributed to exposure to substances hazardous to health, at the earliest possible stage;

(b) to help in assessing the effectiveness of control measures;

(c) the collection, maintenance and use of data for the detection and evaluation of hazards to health;

(d) to assess the immunological status of employees doing specific work with micro-organisms hazardous to health.

Outcome

The results of any health surveillance procedure should lead to some form of action which will be to the benefit of employees.

SUMMARY – CHAPTER 5

1. A substantial number of working days are lost every year as a result of occupational ill-health.

2. Occupational health is concerned with the relationship of work to health and the effects of work on the worker.

3. Occupational hygiene is concerned with the identification, measurement, evaluation and control of contaminants and other physical phenomena which can effect the health of people exposed to them.

4. The causes of occupational diseases and conditions may be of a physical, chemical, biological and work-related nature.

5. Occupational health management is directed at the prevention of occupational ill-health.

6. First aid is an important post-accident strategy.

7. Health surveillance may be required where employees are exposed to health risks at work.

CHAPTER 6

THE WORKING ENVIRONMENT

The design and control of the working environment is an important aspect of health and safety at work. Environmental working conditions in a place of work have a direct effect on human behaviour at work, the degree of risk of occupational disease and/or injury and on morale, management/worker relations, labour turnover and profitability.

Fundamentally, three aspects must be considered:

(a) the organisation of the working environment;

(b) the prevention or control of environmental stressors;

(c) welfare arrangements.

The Workplace (Health, Safety and Welfare) Regulations 1992 replace some 35 pieces of the former legislation – including parts of the Factories Act 1961 and the Offices, shops and Railway Premises Act 1963 – and apply to all places. There are a number of exceptions to the Regulations, namely means of transport, construction sites, sites where extraction of mineral resources or exploration for it is carried out and fishing boats. Workplaces on agricultural or forestry land away from main buildings are also exempted from the requirements.

'Workplace' is broadly defined as meaning any premises or part of premises that are not domestic premises and made available to any person as a place of work, and includes:

(a) any place within the premises to which such person has access while at work;

(b) any room, lobby, corridor, staircase, road or other place used as a means of access to or egress from the workplace other than a public road.

Duty to Maintain the Workplace

There is an absolute duty on employers to maintain the workplace, equipment devices and systems (including cleaning as appropriate) in an efficient state, in efficient working order and in good repair.

The Regulations establish further general duties on employers as outlined below.

The Working Environment

1. Ventilation must be effective and suitable to maintain comfort conditions.

2. The temperature must be reasonable during working hours.

3. Lighting must be suitable in terms of the type of lighting and sufficient with regard to the amount of light provided. Emergency lighting must be provided where failure of lighting may create danger.

4. There is a general duty to keep the workplace clean and surfaces (floors, walls, etc.) must be cleanable. There must be adequate control over refuse.

5. Sufficient floor area and space must be provided for employees.

6. Both indoor and outdoor workstations must be suitable for the work being carried out. Seats provided must be suitable for the person and for the work undertaken.

Structural Safety

1. Floors and traffic routes must be suitable for the purpose, in good repair and free from obstructions.

2. Measures to prevent risks arising from falls and falling objects must be taken, including marking of risk areas and covering of pits, tanks, etc.

3. Windows and transparent or translucent doors, gates and walls must be of safety material or protected in certain cases and be suitably marked.

4. The design of windows, skylights and ventilators must be such that there is no risk to an individual when opening, closing or adjusting and be suitably positioned.

5. Windows etc. must be capable of being cleaned safely, including the fitting suitable devices to buildings.

6. Traffic routes must be organised in such a way as to ensure safe circulation by pedestrians and vehicles.

7. Doors and gates must be suitable designed and constructed and fitted with certain safety devices.

8. Escalators and moving walkways must function safely and safety devices and emergency stop controls must be fitted.

Welfare Amenity Provisions

1. Sanitary conveniences must be suitable and sufficient, adequately ventilated and lit. They must be kept in a clean and orderly condition with

separate facilities for each of the sexes.

2. Washing facilities must be suitable and sufficient, located in the vicinity of sanitary conveniences (and changing rooms if required) with hot and cold water, cleaning and drying facilities. They must be ventilated and well lit and kept in a clean and orderly condition with separate facilities for each of the sexes.

3. Drinking water supply must be wholesome, readily accessible, conspicuously marked with cups or drinking vessels provided unless the supply is from a 'water fountain' facility.

4. Accommodation for clothing must be suitable and sufficient, with security for clothing not worn and separate accommodation in some cases; facilities for drying must be provided which are suitable located.

5. Facilities for changing clothing must be suitable and sufficient and include the provision of rest rooms or areas. There must be separation of smokers and non-smokers, facilities for pregnant women and nursing mothers and facilities must also be provided for eating meals.

Duties under the Regulations are largely of an absolute nature. The Regulations are supported by an ACOP issued by the Health and Safety Commission.

ENVIRONMENTAL STRESSORS

Environmental stress, associated with extremes of temperature, poor standards of lighting and ventilation, noise and vibration, is a contributory factor to accidents. Many forms of occupational ill-health (e.g. heat stroke, hypothermia, noise-induced hearing loss) are also directly associated with a failure to provide a healthy working environment. The various forms of environmental stressor are dealt with below.

Temperature

Generally, the temperature in a workplace should be 'reasonable'. This implies the need to consider the type of work undertaken (e.g. active or sedentary work, the season of the year, etc.). Temperatures in workrooms should normally be at least 16°C unless much of the work involves severe physical effort in which case the temperature should be at least 13°C.

Ventilation

Two aspects must be considered here, namely:

(a) the provision of adequate 'comfort' ventilation in terms sufficient quantity of fresh or purified air;

(b) means for the removal of airborne contaminants (e.g. dusts, gases, vapours, fumes, etc.) by the operation and maintenance of *local exhaust ventilation* (LEV) systems.

Humidity

Excessive amounts of moisture from, for instance, steam producing equipment, can have a direct effect on comfort. Relative humidity should be between 30 per cent and 70 per cent. If the relative humidity is too low, a feeling of discomfort is produced due to drying of the throat and nasal passages. Conversely, high levels of relative humidity produce a sensation of stuffiness and reduces the rate at which sweat evaporates.

Lighting

Every workplace must have 'suitable and sufficient' lighting: 'suitable' in terms of qualitative factors, such as freedom from glare, correct distribution, etc. and 'sufficient' as far as the actual amount of light, measured in Lux, for both general areas and specific tasks.

Noise and Vibration

Exposure to noise above 90 dB can cause noise-induced hearing loss. Excessive noise can distract attention, affect concentration, mask audible warning signals or interfere with the work process, thereby becoming a contributory factor to accidents.

Many cases of work-related upper limb disorders are associated with the use of various rotary and percussive hand tools. Exposure to whole body vibration associated, for instance, with driving heavy lorries long distances, can cause blurred vision, loss of balance and loss of concentration in some cases.

WELFARE AMENITY PROVISIONS: FACTORS FOR CONSIDERATION

In the design of welfare amenity provisions, the following points should be considered.

Design Aspects

Surfaces; intervening ventilated space between sanitation area and workrooms; disposal of sanitary dressings; lighting and ventilation; vandalproof fittings

AREA	FREQUENCY	RESPONSE	EQUIPMENT	MATERIALS	METHOD	SPECIAL PRECAUTIONS
Yard	Daily		Hose and broom		Sweep up all refuse hose and brush down yard and surfaces.	Keep drain gullies clear.
Floor Stores Preparation Kitchen, Wash-Up Servery, Lavatory, Wash Room	Daily		Vacuum cleaner suction polisher, polisher floor scrubber, floor drier, sponge map. Plastic bucket.	Detergent/ sanitiser warm water	Vacuum clean wash floors, dry floor. Policy (where appropriate).	
Walls All rooms as above	Fortnightly Monthly		Vacuum cleaner (with attachment). Mechanical wall washer or sponge and plastic bucket.	Detergent/ sanitiser warm water	Vacuum clean. Wash down.	Ventilation fans should be included in the cleaning, but the electric supply to the fan must be switched off at the main. Include canopies over cooking equipment.
Ceilings	Monthly 6 monthly		Vacuum cleaner (with attachment) Sponge and plastic bucket.	Detergent/ sanitiser water water	Vacuum clean. Wash ceilings.	
Lighting Windows Interior Exterior Electric fittings	Fortnightly Monthly Monthly		Chamois leather plastic bucket. Drying cloth plastic bucket.	Water	Wash.	Safe working conditions must be ensured. Ladders must be sound and firmly placed. The electricity supply should be switched off at the mains.
WCs Pans	Daily		Nylon brush.	Approved cleanser	Sprinkle cleanser around bowl at end of each day. Brush entire bowl surface following morning. Flush pan, holding brush under flush water.	When an approved cleanser is used, no other cleansing agent must be mixed with it or used at the same time.
Seat Flushing	Daily		Disposable cloth plastic bucket.	Detergent/ sanitiser	Thoroughly wash the upper and lower sides of the seat, the door, furniture and the flushing handle. Dispose of the cloth via the WC pan.	
Wash hand basins/ showers	Daily		Sponge.	Approved cleanser	Thoroughly wash the basin.	

and surfaces; adequate hot and cold water supplies; soap, nailbrushes, means of drying hands; seats for changing footwear: means of drying clothing; drinking water fountains.

Location

Ease of access and proximity to the workplace; main location or series of subsidiary locations.

Layout

Separate access to clothing storage, sanitation, hand cleansing and shower facilities; no overcrowding; sequential flow arrangement; clothing storage rails; lockers for personal possessions etc.; siting of wash basins, showers on the basis of 1:40 recommended standard.

SUMMARY – CHAPTER 6

1. The working environment has a direct effect on the potential for accidents and occupational ill-health.

2. The principal areas for consideration are the organisation of the working environment, including structural safety aspects, the prevention of environmental stressors and the arrangements for welfare.

3. Environmental stressors include inadequate temperature and humidity control, inadequate lighting and ventilation, noise and vibration.

4. The design of welfare arrangements and the control of environmental stressors are incorporated in the Workplace (Health, Safety and Welfare) Regulations 1992 and ACOP.

CHAPTER 7

HAZARDOUS SUBSTANCES

CLASSIFICATION OF HAZARDOUS SUBSTANCES

Hazardous substances are classified according to Schedule 1 of the Chemicals (Hazard Information and Packaging for Supply) [CHIP 2] Regulations 1994 as outlined below.

Part I
Categories of Danger

Category of danger	Property (see note 1)	Symbol or Letter
Physico-chemical Properties		
Explosive	Solid, liquid, pasty or gelatinous substances and preparations which may react exothermically without atmospheric oxygen thereby quickly evolving gases and which under defined test conditions detonate, quickly deflagrate or upon heating explode when partially confined.	E
Oxidising	Substances and preparations which give rise to a exothermic reaction on contact with other substances, particularly flammable substances.	O
Extremely Flammable	Liquid substances and preparations having and extremely low flash point and a low boiling point. Gaseous substances and preparations that are flammable in contact with air at ambient temperatures and pressures.	F+
Highly Flammable	The following substances and preparations: (a) that may become hot and finally catch fire on contact with air at ambient temperatures without any application of energy;	

Category of danger	Property (see note 1)	Symbol or Letter
	(b) that may readily catch fire after brief contact with a source of ignition and that continue to burn or to be consumed after the source of ignition has been removed;	
	(c) that have a very low flash point;	
	(d) that, in contact with water or damp air, evolve highly flammable gases in dangerous quantities. (See note 2.)	
Flammable	Liquid substances and preparations having a low flash point.	F

Health Effects

Very Toxic	Substances and preparations that in very low quantities can cause death or acute or chronic damage to health when inhaled, swallowed or absorbed via the skin.	T+
Toxic	Substances and preparations that in low quantities can cause death or acute or chronic damage to health when inhaled, swallowed or absorbed via the skin.	T
Harmful	Substances and preparations that may cause death or acute or chronic damage to health when inhaled, swallowed or absorbed via the skin.	Xn
Corrosive	Substances and preparations that may, on contact with living tissues, destroy them.	C
Irritant	Non-corrosive substances and preparations that through immediate, prolonged or repeated contact with the skin or mucous membrane may cause inflammation.	Xi
Sensitising	Substances and preparations that, if they are inhaled or if they penetrate the skin, are capable of eliciting a reaction by hypersensitisation such that on further exposure to the substance or preparation, characteristic adverse effects are produced.	

Category of danger	Property (see note 1)	Symbol or Letter
Toxic for reproduction (see note 3)	Substances and preparations which, if they are inhaled or ingested or if they penetrate the skin, may produce or increase the incidence of non-heritable adverse effects in the unborn child and/or an impairment of male or female reproductive functions or capacity	
Category 1		T
Category 2		T
Category 3		Xn
Dangerous for the Environment (see note 4)	Substances that, were they to enter into the environment, would present or might present and immediate or delayed danger for one or more components of the environment.	N

Notes

1. As further described in the approved classification and labelling guide.

2. Preparations packed in aerosol dispensers will be classified as flammable in accordance with the additional criteria set out in Part II of the Schedule.

3. The categories are specified in the approved classification and labelling guide.

4. (a) In certain cases specified in the approved supply list and in the approved classification and labelling guide substances classified as dangerous for the environment do not require to be labelled with the symbol for this category of danger.

   (b) This category of danger does not apply to preparations.

Figure 14: Hazard Warning Symbols

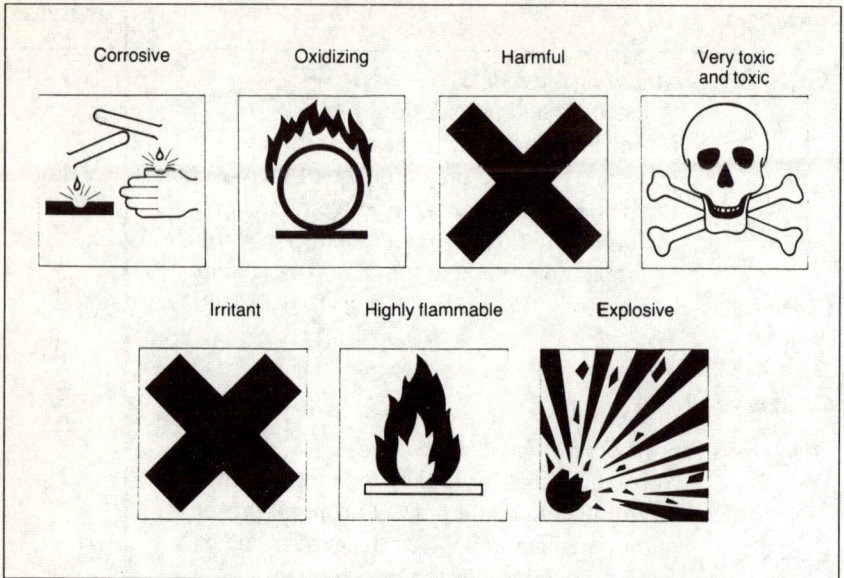

PRINCIPLES OF TOXICOLOGY

Toxicology is the study of the body's responses to toxic substances. *Toxicity*, on the other hand, is related to the ability of a chemical molecule to produce injury once it reaches a susceptible site in or on the body.

Important Terms

Acute effect	A rapidly produced effect following a single exposure to an offending agent.
Chronic effect	An effect produced as a result of prolonged exposure or repeated exposures of long duration. Concentrations of the offending agent may be low in both cases.
Sub-acute effect	A reduced form of acute effect.
Progressive chronic effect	An effect which continues to develop after exposure ceases.
Local effect	An effect usually confined to the initial point of contact. The site may be the skin, mucous membranes of the eyes, nose or throat, liver, bladder, etc.
Systemic effect	Such effects occur in parts of the body other than at the initial point of contact, and are associated with

a particular body system, e.g. respiratory system, central nervous system.

Toxic Substances: routes of entry

Inhalation

Inhalation of toxic substances, in the form of a dust, gas, mist, fog, fume or vapour, accounts for approximately 90 per cent of all ill-health associated with toxic substances. The results may be acute (immediate) as in the case of gassing accidents, e.g. chlorine, carbon monoxide, or chronic (prolonged, cumulative) as in the case of exposure to chlorinated hydrocarbons, lead compounds, benzene, numerous dusts, which produce pneumoconiosis, mists and fogs, such as that from paint spray, oil mist, and fume, such as that from welding operations.

Pervasion

The skin, if intact, is proof against most, but not all, inputs. There are certain substances and micro-organisms which are capable of passing straight through the intact skin into underlying tissue, or even into the bloodstream, without apparently causing any changes in the skin.

The resistance of the skin to external irritants varies with age, sex, race, colour and, to a certain extent, diet. Pervasion, as a route of entry, is normally associated with occupational dermatitis, the causes of which may be broadly divided into two groups:

(a) primary irritants: are substances which will cause dermatitis at the site of contact if permitted to act for a sufficient length of time and in sufficient concentrations, e.g. strong alkalis, acids and solvents;

(b) secondary cutaneous sensitisers: are substances which do not necessarily cause skin changes on first contact, but produce a specific sensitisation of the skin. If further contact occurs after an interval of, say, seven days or more, dermatitis will develop at the site of the second contact. Typical skin sensitisers are plants, rubber, nickel and many chemicals.

It should be noted that, for certain people, dermatitis may be a manifestation of psychological stress, having no relationship with exposure to toxic substances (an endogenous response).

Ingestion

Certain substances are carried into the gut from which some will pass into the body by absorption. Like the lung, the gut behaves as a selective filter which keeps out many but not all harmful agents presented to it.

Figure 15: Dose-Response Relationship

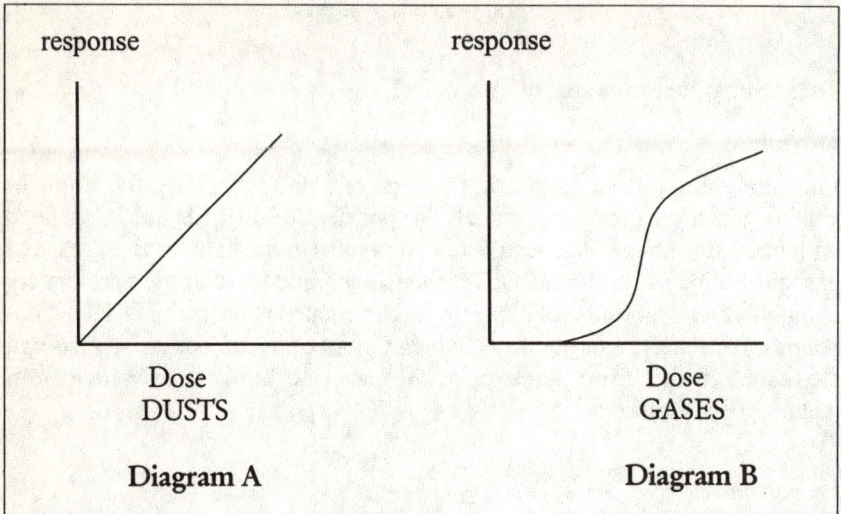

response

Dose
DUSTS

Diagram A

response

Dose
GASES

Diagram B

Injection/Implantation

A forceful breach of the skin, frequently as a cause of injury, can carry substances through the skin barrier.

Toxic substances are widely used in industry. Some indication of the most common hazards and the occupations associated with them are listed in the Social Security (Industrial Injuries) (Prescribed Diseases) Regulations 1985.

The Responses of the Body to Toxic Substances

Dose-Response Relationship

A basic principle of occupational disease prevention rests upon the reality of threshold levels of exposure for the various hazardous agents below which people can cope successfully without a significant threat to their health. This concept derives from the quantitative characteristic of the dose-response relationship, according to which there is a systematic down change in the magnitude of people's response as the dose of the offending agent is reduced.

Dose   =   level of environmental   x   duration of
contamination   exposure

With many dusts, for instance, the body's response is directly proportional to the dose received over a period of time, the greater the dose, the more serious the condition, and vice versa.

However, in the case of airborne contaminants, such as gases or mists, there is a concentration in air or dose below which most people can cope reasonably well. Once this concentration in air is reached (threshold dose) some form of body response will result. This concept is most important in the correct use and interpretation of 'Occupational Exposure Limits' (formerly known as 'Threshold Limit Values').

In Figure 15 above, Diagram A shows a typical direct dose-response relationship, whereas in Diagram B the dose-response curve reaches a level of "no response" at a point greater than zero on the dose axis. This point of cut-off identifies the threshold dose which was the original basis for the setting of 'Occupational Exposure Limits'.

Target Organs and Systems

It is a well-established fact that certain substances have a direct or indirect effect on certain body organs (target organs) and certain body systems (target systems). Target organs include the lungs, liver, brain, skin and bladder. Target systems, on the other hand, include the central nervous system, circulatory system and the reproductive system.

PROTECTIVE MECHANISMS

The protective mechanisms within the body respond largely according to the shape and size of particulate matter which may be inhaled. Such mechanisms are as follows.

Nose   The coarse hairs in the nose, assisted by mucus from the nasal lining, act as a filter, trapping the larger particles of dust.

Ciliary escalator   The respiratory tract consists of the trachea (windpipe) and bronchi which branch to the lungs. The lining of the trachea consists of quite tall cells, each of which has a cilium growing from its head (ciliated epithelium). These cilia exhibit a wave-like motion so that a particle falling onto the cilia is returned back to the throat. Mucus assists these particles to stick.

Macrophages   These are wandering scavenger cells. They have an irregular outline and large nucleus, and can move freely through body tissue, engulfing bacteria and dust particles. They secrete hydrolytic enzymes which attack any foreign bodies.

Lymphatic system   This is a drainage system which acts as a clearance channel for the removal of foreign bodies, many of which are retained in the lymph nodes throughout the body. In certain cases, a localised inflammation will be set up in the lymph node.

Tissue response   A typical example of tissue response is in byssinosis (the "Monday fever" or "Monday feeling"), a chest condition of cotton workers, where the lung becomes sensitised to cotton dust through continuing exposure.

SAFE HANDLING OF TOXIC SUBSTANCES

Assuming that substitution as a control strategy has been considered and found impracticable, then in order of merit:

(a) use in diluted form wherever possible;

(b) only limited quantities should be used or stored at any one time; large quantities should be stored in a purpose-built bulk chemical store;

(c) containment of the specific area; consider safe venting and drainage requirements;

(d) eliminate handling and dispensing from bulk; use of automatic systems, e.g. cleaning-in-place (CIP) systems;

(e) provide adequate exhaust ventilation;

(f) separation, e.g. acids from alkalis;

(g) personal protective equipment as an extra, not sole, means of protection.

THE PHYSICAL STATE OF HAZARDOUS SUBSTANCES

In the evaluation of the risks associated with hazardous substances and any air monitoring to be undertaken, it is necessary to consider the actual form taken by that substance. Substances carried in air are *aerosols*.

Forms of Aerosol

Features of the various forms of aerosol are outlined below.

Dusts   The International Labour Organisation defines dust as an aerosol composed of solid inanimate particles. Dusts are solid airborne particles, often created by operations such as grinding, crushing, milling, sanding and demolition. Two of the principal dusts encountered in industry are asbestos and silica.

Dusts may be	(a) fibrogenic: they cause fibrotic changes to lung tissue, e.g. silica, cement dust, coal dust and certain metals; (b) toxic: they eventually poison the body systems, e.g. arsenic, mercury, beryllium, phosphorus and lead.
Mists	A mist comprises airborne liquid droplets, a finely dispersed liquid suspended in air. Mists are mainly created by spraying, foaming, pickling and electro-plating. Danger arises most frequently from acid mist produced in industrial treatment processes, e.g. oil mist or chromic acid mist.
Fumes	These are fine solid particulates formed from the gaseous state usually by vaporisation or oxidation of metals, e.g. lead fumes. Fumes usually form an oxide in contact with air. They are created by industrial processes which involve the heating and melting of metals, such as welding, smelting and arc air gouging. A common fume danger is lead poisoning associated with the inhalation of lead fumes.
Gases	These are formless fluids usually produced by chemical processes involving combustion or by the interaction of chemical substances. A gas will normally seek to completely fill the space into which it is liberated. A classic gas encountered in industry is carbon monoxide. Certain gases such as acetylene, hydrogen and methane are particularly flammable.
Vapours	A vapour is the gaseous form of a material normally encountered in a solid or liquid state at normal room temperature and pressure. Typical examples are solvents, such as trichloroethylene, which release vapours when the container is opened. Other liquids produce a vapour on heating, the amount of vapour being directly related to the boiling point of that particular liquid. A vapour contains very minute droplets of the liquid. However, in the case of a fog, the liquid droplets are much larger.
Smoke	Smoke is a product of incomplete combustion, mainly of organic materials. It may include fine particles of carbon in the form of ash, soot and grit that are visibly suspended in air.

PREVENTION AND CONTROL STRATEGIES

The more important prevention and control strategies, in order of effectiveness, are: prohibition, elimination, substitution, enclosure/containment, isolation/separation, local exhaust ventilation, dilution ventilation with personal protective equipment as the last resort.

PREVENTION STRATEGIES

Prohibition

In certain cases a substance may be so inherently dangerous that its use may be prohibited by law or organisation policy.

Elimination

Review of the needs of specific processes can often reveal chemicals and processes which are no longer necessary. Where such chemicals can be eliminated from use, control is unnecessary.

Substitution

Can a safer material be used? There are some substances which should never be used and are prescribed in current legislation. Others may be banned as a matter of policy within organisations. Typical examples are the substitution of toluene for benzene, fibreglass for asbestos, or trichloroethane for trichloroethylene.

CONTROL STRATEGIES

Enclosure/Containment

Can the materials be handled so that individuals never need come into contact with them? Total enclosure or containment of the process may be possible by the use of bulk tanks and pipework to deliver a liquid directly into a closed production vessel. Complete enclosure is practicable if the substances are in liquid form, used in large quantities, and if the range of substances is small.

Isolation/Separation

Can the process be put somewhere else? The isolation of a process may simply mean putting it into a small locked room, thereby separating the workforce from the risk, or could involve the construction of a chemical plant in a remote geographical area. The system of isolation is required to prevent access effectively, or certainly restrict access only to those who need to be there.

Ventilation systems

Ventilation is an important control strategy. Here it is necessary to distinguish between natural ventilation and mechanical ventilation systems.

Air may enter a building by one or more of the following means:

Infiltration

Infiltration through the fabric of a building, for example through gaps around doors and windows or between roofing panels or tiles, is common. Many traditional buildings have infiltration rates of between 0.5 and 2 air changes per hour.

Planned Natural Ventilation

This takes place through fixed openings or vents, or through windows and doors.

Dilution Ventilation

In certain cases, it may not be possible to extract a contaminant close to the point of origin. If the quantity of contaminant is small, uniformly evolved

Figure 16: Local Exhaust Ventilation Systems

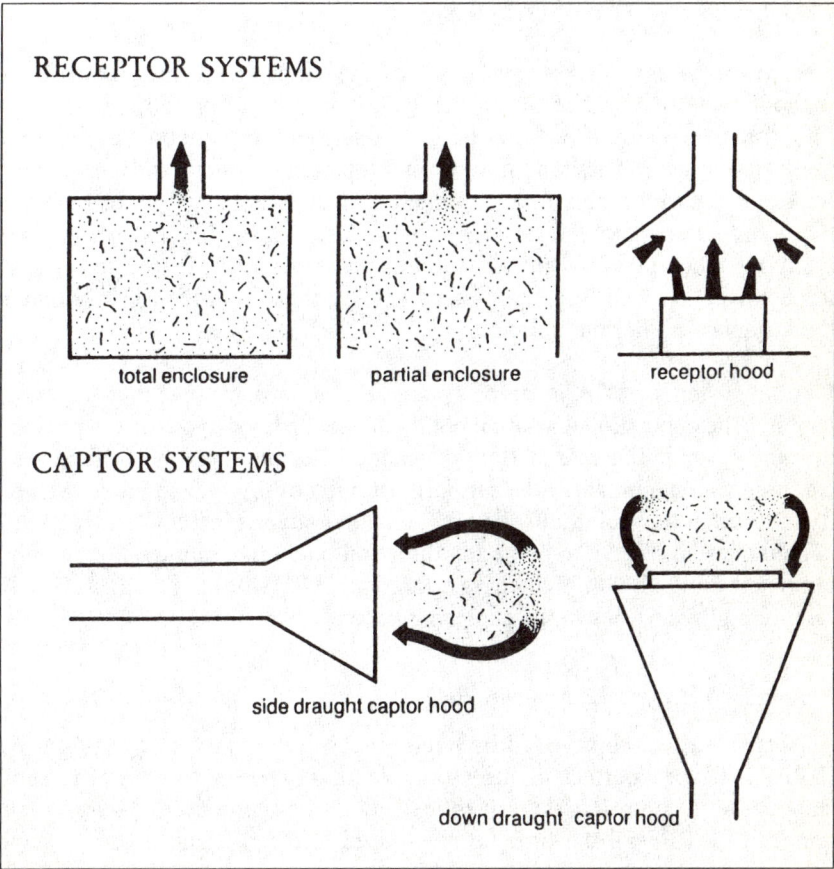

RECEPTOR SYSTEMS

total enclosure   partial enclosure   receptor hood

CAPTOR SYSTEMS

side draught captor hood

down draught captor hood

and of low toxicity, it may be possible to dilute the contaminant by inducing large volumes of air to flow through the contaminated region. Dilution ventilation is most successfully used to control vapours, for example, organic vapours from low toxicity solvents, but is seldom successfully applied to dust and fumes, as it will not prevent inhalation.

Mechanical Ventilation

This operates either through a system of extractor fans located in the wall and/or roof of a building, or by means of more complex ducted ventilation systems designed to remove the contaminant at the point of emission (Local Exhaust Ventilation – LEV). In certain cases dilution ventilation may be appropriate, or a system of air conditioning may operate.

Infiltration and planned natural ventilation give no continuing protection wherever toxic gases, fumes, vapours, etc. are emitted from processes. LEV systems must, therefore, be operated. These take two principal forms, receptor systems and captor systems and examples of them are shown in Fig. 16, above.

LEV Systems

Receptor systems   In a receptor system, the contaminant enters the system without inducement. The fan in the system is used to provide air flow to transport the contaminant from the hood/enclosure through the ducting to a collection system. The hood may form a total enclosure around the source, for example, with highly toxic contaminants, such as beryllium or radioactive sources, or a partial enclosure, as in the case of spray booths, in which all spraying takes place within the booth, and laboratory fume cupboards. Receptor hoods receive contaminants as they flow from their origin under the influence of thermal currents.

Captor systems   With a captor system, the air which flows into the hood captures the contaminant at some point outside the hood and induces its flow into the system. The rate of flow of air into the hood must be sufficient to capture the contaminant at the furthermost point of origin, and the air velocity induced at this point must be high enough to overcome any tendency the contaminant may have to go in any direction other than into the hood. Contaminants emitted with high energy, large particles with high velocities will require high velocities in the capturing stream.

Personal Protective Equipment (PPE)

The use of various forms of PPE, including respiratory protective equipment (RPE), is never a perfect solution to preventing exposure to hazardous substances. As a control strategy it relies heavily on the operator wearing the correct PPE/RPE all the time he is exposed to the risk and people do not always do this.

In the majority of the cases the provision and use of PPE should be seen as an extra form of protection where other forms of protection, as indicated above, are operating.

Figure 17: Form of Dangerous Substances Audit

DANGEROUS SUBSTANCES AUDIT

	YES/NO	ACTION
1. INFORMATION AND IDENTIFICATION		
1.1 Is an up-to-date list of all substances hazardous to health used and stored on site readily available?		
1.2 Are safety data sheets available for all substances hazardous to health on site?		
1.3 Is the information adequate in each case?		
1.4 Are all packages and containers correctly labelled?		
1.5 Are 'ready use' containers suitable for that purpose and suitably marked?		
2. STORAGE		
2.1 Are internal stores and external storage areas satisfactory in respect of construction, layout, security and control?		
2.2 Are substances hazardous to health correctly segregated from other substances?		
2.3 Are cleaning and housekeeping levels satisfactory in storage areas?		
2.4 Are all substances issued to personnel controlled?		
3. PROTECTION		
3.1 Are the necessary safety signs displayed prominently in the appropriate areas?		
3.2 Is suitable personal protective equipment: (a) available? (b) serviceable? (c) used/worn?		
3.3 Are emergency shower and eye wash facilities: (a) available? (b) suitably located? (c) serviceable?		
3.4 Are the above facilities frost protected?		
3.5 Are adequate and suitable first aid facilities:		

	YES/NO	ACTION

(a) available?
(b) suitably located?
3.6 Are the appropriate fire appliances:
    (a) available?
    (b) suitably located?
    (c) serviceable?
    (d) accessible?
3.7 Is a supply of neutralising compound available in the event of spillage?

4. PROCEDURES
4.1 Are written procedures for safe handling available for all substances hazardous to health?
4.2 Is a written spillage procedure available?
4.3 Is there a routine inspection procedure in respect of:
    (a) personal protective equipment?
    (b) emergency shower and eye wash facilities?
    (c) first aid boxes?
    (d) fire appliances?
    (e) chemical dosing to equipment?
    (f) neutralising compounds?

5. TRAINING
5.1 Are staff trained in:
    (a) safe handling procedures?
    (b) use of fire appliances?
    (c) dealing with spillages?
    (d) use and care of personal protective equipment?
5.2 Are training records maintained?

ACTION PROGRAMME

1. Immediate action

2. Short term action (7 days)

3. Medium term action (1 month)

4. Long term action (1 year)

Auditor _____   Date _____

SUMMARY – CHAPTER 7

1. All substances must be classified according to Schedule 1 of the CHIP Regulations.

2. Toxicology is the study of the body's responses to toxic substances.

3. The effects of exposure to toxic substances may be acute, chronic, local or systemic.

4. In evaluating the risks arising from exposure to hazardous substances, the route of entry is significant.

5. The principal route of entry to the body of hazardous substances is through inhalation of a gas, vapour, mist or other form of aerosol.

6. The concept of dose-response relationship is significant in assessing health risks from hazardous substances.

7. Certain substances attack identified target organs and/or systems.

8. A strategy must be devised to ensure the safe handling of toxic and other hazardous substances.

9. The physical state of a hazardous substance is significant in its potential for harm.

10. Risk assessment of hazardous substances should incorporate means for preventing or controlling exposure.

11. The design of LEV systems to control exposure to aerosols is of prime importance.

PART 3

HUMAN FACTORS AND SAFETY

CHAPTER 8

INTRODUCTION TO HUMAN FACTORS

Current health and safety legislation requires a human factors-related approach to occupational health and safety. Employers must consider human capability from a health and safety viewpoint when entrusting tasks to their employees.

But what is meant by 'human capability'? Various terms are found in the average dictionary: 'able', 'competent', 'gifted' and 'having the capacity'. Perhaps the last term is the most significant from a health and safety viewpoint. 'Capacity' implies both mental and physical capacity, for instance, the mental capacity to understand why a task should be undertaken in a particular way, and physical capacity, in terms of the actual physical strength and fitness to undertake the task in question.

WHAT ARE 'HUMAN FACTORS'?

'Human factors', in their application to occupational health and safety, have been defined as "a range of issues including the perceptual, physical and mental capabilities of people and the interactions of individuals with their jobs and working environments, the influence of equipment and system design on human performance and those organisational characteristics which influence safety-related behaviour at work" (HSE).

Although most health and safety legislation places the duty of compliance firmly on the employer or body corporate, this duty can only be discharged by the effective actions of its managers. For instance, it is management's job to report and investigate accidents at work, but how frequently the causes are written down to "operator carelessness", "not looking where he was going" or, quite simply, "human error", indicating that nothing can be done and no further action should be taken.

AREAS OF INFLUENCE ON PEOPLE AT WORK

There is no doubt that people are influenced by a range of factors at work. There are, however, broadly three such areas of influence: the organisation, the job and personal factors. These areas are directly affected by the system for communication within the organisation and the training systems and procedures in operation, all of which are directed at preventing human error. These factors are considered below.

The Organisation

Those organisational characteristics which influence safety-related behaviour include:

(a) the need to promote a positive climate in which health and safety is seen by both management and employees as being fundamental to the organisation's day-to-day operations, i.e. they must create a positive safety culture;

(b) the need to ensure that policies and systems which are devised for the control of risk from the organisation's operations take proper account of human capabilities and fallibilities;

(c) commitment to the achievement of progressively higher standards which shown at the top of the organisation and cascaded through successive levels;

(d) demonstration by senior management of their active involvement, thereby galvanizing managers throughout the organisation into action;

(e) leadership, whereby an environment is created which encourages safe behaviour.

The Job

Successful management of human factors and the control of risk involves the development of systems of work designed to take account of human capabilities and fallibilities. Using techniques like job safety analysis, jobs should be designed in accordance with ergonomic principles so as to take into account limitations in human performance.

Major considerations in job design include:

(a) identification and comprehensive analysis of critical tasks expected of individuals and appraisal of likely errors;

(b) evaluation of required operator decision-making and the optimum balance between the human and automatic contributions to safety actions;

(c) application of ergonomic principles to the design of man-machine interfaces, including displays of plant and process information, control devices and panel layouts;

(d) design and presentation of procedures and operating instructions;

(e) organisation and control of the working environment, including workspace, access for maintenance, noise, lighting and thermal conditions;

(f) provision of correct tools and equipment;

(g) scheduling of work patterns, including shift organisation, control of fatigue and stress and arrangements for emergency operations;

(h) efficient communications, both immediate and over periods of time.

Personal Factors

This aspect is concerned with how personal factors such as attitude, motivation, training, human error and the perceptual, physical and mental capabilities of people can interact with health and safety issues.

Attitudes are directly connected with an individual's self-image, the influence of groups and the need to comply with group norms or standards and, to some extent, opinions, including superstitions (like "all accidents are acts of God"). Changing attitudes is difficult. They may be formed through past experience, the level of intelligence of the individual, specific motivation, financial gain and skills available to an individual. There is no doubt that management example is the strongest of all motivators to bring about attitude change.

Important factors in motivating people to work safely include joint consultation in planning the work organisation, the use of working parties or committees to define objectives, attitudes currently held, the system for communication within the organisation and the quality of leadership at all levels. Financially-related motivation schemes, such as safety bonuses, do not necessarily change attitudes, people frequently reverting to normal behaviour when the bonus scheme finishes.

HUMAN ERROR

Is human error a significant feature of accidents? How often have we seen the cause of an accident written down to "carelessness" on the part of the individual. The fact is that whilst people may occasionally be genuinely careless, in most cases the cause is more likely to be human error.

Limitations in human capacity to perceive, to attend to, to remember, to process and to act on information are all relevant in the context of human error.

Typical human errors are associated with lapses of attention, mistaken actions, mis-perceptions, mistaken priorities and, in a limited number of cases, wilfulness.

SUMMARY – CHAPTER 8

1. Employers must consider human capability when entrusting tasks to their employees.

2. Human factors are concerned with the organisation, the job and personal factors.

3. People are influenced by communication and training systems.

4. Consideration of those organisational characteristics which influence safety-related behaviour is important.

5. Systems of work must take account of human capabilities and fallibilities.

6. Job design must consider, in particular, critical tasks, individual decision making requirements, the application of ergonomic principles and the need for efficient communication.

7. People do make mistakes. Employers must consider the potential for human error when designing jobs and allocating tasks.

8. Personal factors, such as attitudes held and motivating factors, are significant.

CHAPTER 9

ERGONOMICS

Ergonomics (the scientific study of work) is a branch of science which involves the consideration of people at work in terms of human characteristics, the tasks they undertake, the machinery they operate (the "man-machine interface"), the environments within which they work and how these various factors affect the total working system. More appropriate definitions are "human factors engineering" or "fitting the task to the individual".

As such, ergonomics embraces a number of disciplines including physiology, anatomy, psychology, engineering and environmental science. It examines, in particular, the physical and mental capacities and limitations of workers taking into account, at the same time, psychological factors (such as learning, individual skills, perception, attitudes, vigilance, information processing and memory) together with physical factors (such as strength, stamina and body dimensions).

Fundamentally, ergonomics is concerned with maximizing human performance and, at the same time, eliminating, as far as possible, the potential for human error.

MAIN AREAS FOR CONSIDERATION

The human system   People are different in terms of physical and mental capacity. This is particularly apparent when considering the physical elements of body dimensions, strength and stamina, coupled with the psychological elements of learning, perception, personality, attitude, motivation and reactions to given stimuli.

Other factors which have a direct effect on performance include the level of knowledge and the degree of training received, a person's own personal skills and experience of the work.

Environmental factors   The consideration of the working environment in terms of layout of the working area and the amount of individual work space available, together with the need to eliminate or control environmental stressors is an important feature of ergonomic consideration. Environmental stressors, which have a direct effect on health of people at work and their subsequent performance, include noise, vibration, extremes of temperature and humidity, poor levels of lighting and ventilation.

The man-machine interface   Machines are designed to provide information to operators through various forms of display, such as a temperature gauge. Similarly, the operator must ensure the correct and safe operation of the machine through a system of controls, such as push-button *STOP* and *START* controls, foot pedals or manual devices. The study of displays, controls and other design features of vehicles, machinery, automation and communication systems, with a view to reducing operator error and stress on the operator, is a significant feature of design ergonomics.

Factors such as the location, reliability, ease of operation and distinction of controls and the identification, ease of reading, sufficiency, meaning and compatibility of displays are all significant in ensuring correct operation of the various forms of work equipment.

The total working system   Considering the above factors, the approach to ergonomics can be summarized under the heading of *The Total Working System*, which is broken down into four major elements as shown below.

THE TOTAL WORKING SYSTEM

Human characteristics	**Environmental factors**
Body dimensions	Temperature
Strength	Humidity
Physical and mental limitations	Lighting
Stamina	Ventilation
Learning	Noise
Perception	Vibration
Reaction	Airborne contamination
Man-machine interface	**Total Working System**
Displays	Productivity
Controls	Accidents and ill health
Communications	Health and safety protection
Automation	Fatigue
	Posture
	Work rate

ANTHROPOMETRY

A key feature of ergonomic design is that of matching people to the equipment they use at work. Anthropometry is the study and measurement of body
dimensions, the orderly treatment of resulting data and the application of
those data in the design of workspace layouts and equipment.

Few workstations are 'made to measure' owing to the wide range of human dimensions and the sheer cost of designing individual workstations and
machines to conform with individual body requirements. The fact that this is
not done creates many problems, best demonstrated by research at the
Cranfield Institute of Technology, who created *Cranfield Man.*

Cranfield Man

Figure 18: Cranfield Man – 1.35m tall with a 2.44m arm span

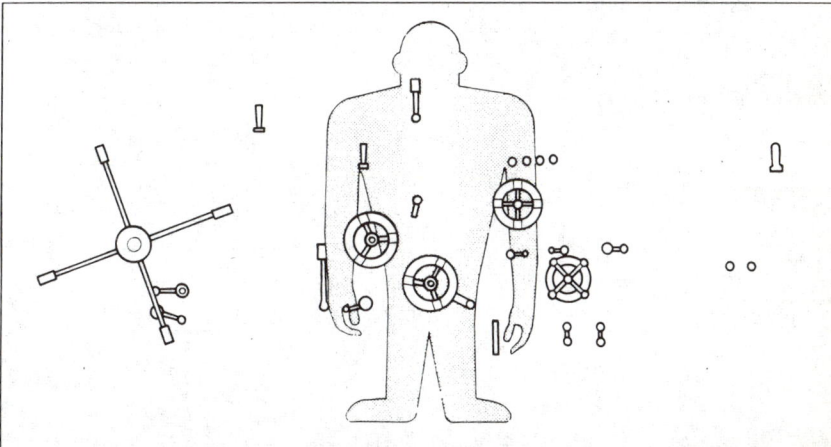

Using a horizontal lathe, researchers examined the positions of controls and
compared the locations of these controls with the physical dimensions of the
average operator. The table below shows the differences between the two.

**Table 2: Physical dimensions of the average operator compared with
those of *Cranfield Man***

Average Operator	Dimensions	Cranfield Man
1.75m	Height	1.35m
0.48m	Shoulder width	0.61m
1.83m	Arm span	2.44m
1.07m	Elbow height	0.76m

SUMMARY – CHAPTER 9

1. Ergonomics is variously defined as "the scientific study of work", "the study of the man-machine interface", "human engineering" and "fitting the task to the individual".

2. Ergonomics is concerned with maximizing human performance and, at the same time, eliminating the potential for human error.

3. The four areas of ergonomic study are: the human system, environmental factors, the man-machine interface and the total working system.

4. Anthropometry, a branch of ergonomics, is the study and measurement of body dimensions, the orderly treatment of the resulting data, and the application of those data in the design of workspace layouts and equipment.

CHAPTER 10

MANUAL HANDLING OPERATIONS

Typical injuries and conditions associated with manual handling can be both external and internal. External injuries include cuts, bruises, crush injuries and lacerations to fingers, forearms, ankles and feet. Generally, such injuries are not as serious as the internal forms of injury which include muscle and ligamental tears, hernias (ruptures), prolapsed intervertebral discs (slipped discs) and damage to knee, shoulder and elbow joints.

Muscle and Ligamental Strain

Muscle is the most abundant tissue in the body, and accounts for some two-fifths of the body weight. The specialized component is the muscle fibre, a long slender cell or agglomeration of cells which become shorter and thicker in response to a stimulus. The fibres are supported and bound by ordinary connective tissue, and are well supplied with blood vessels and nerves. When muscles are utilised for manual handling purposes, they are subjected to varying degrees of stress. Carrying generally imposes a pronounced static strain of many groups of muscles, especially those of the arms and the trunk. This is a particularly unsuitable form of work for human beings because the blood vessels in the contracted muscles are compressed and the flow of blood, and with it, the oxygen and sugar supply, is thereby impeded. As a result, fatigue very soon sets in, with pains in the back muscles, which perform static work only, occurring sooner than in the arm muscles, which perform essentially dynamic work.

Ligaments are fibrous bands occurring between two bones at a joint. They are flexible but inelastic, come into play only at the extremes of movement, and cannot be stretched when they are taut. Ligaments set the limits beyond which no movement is possible in a joint. A joint can be forced beyond its normal range only by tearing a ligament: this is a sprain. Fibrous tissue heals reluctantly, and a severe sprain can be as incapacitating as a fracture. There are many causes of torn ligament, in particular, jerky handling movements which place stress on a joint, unco-ordinated team lifting, and dropping a load half-way through a lift, often caused by failing to assess the load prior to lifting.

Hernia (Rupture)

A hernia is a protrusion of an organ from one compartment of the body into another, e.g. a loop of intestine into the groin or through the frontal abdominal wall. A hernia can result from incorrect handling techniques and particularly from the adoption of bent back stances, which produce compression of the abdomen and lower intestines.

The most common form of hernia (or 'rupture') associated with manual handling is the inguinal hernia. The weak point is the small gap in the abdominal muscles where the testis descends to the scrotum. Its vessels pass through the gap, which therefore cannot be sealed. Excessive straining, and even coughing, may cause a bulge at the gap and a loop of intestine or other abdominal structure easily slips into it. An inguinal hernia sometimes causes little trouble, but it can, without warning, become strangulated, whereby the loop of intestine is pinched at the entrance to the hernia. Its contents are obstructed and fresh blood no longer reaches the area. Prompt attention is needed to preserve the patient's health, and even his life may be at risk if the condition does not receive swift attention. The defect, in most cases, must be repaired surgically.

Prolapsed or 'slipped' Disc

The spine consists of a number of small interlocking bones or vertebrae. There are seven neck or cervical vertebrae, twelve thoracic vertebrae, five lumbar vertebrae, five sacral vertebrae and four caudal vertebrae. The sacral vertebrae are united, as are the caudal vertebrae, the others being capable of independent but co-ordinating articulating movement. Each vertebra is separated from the next by a pad of gristle-like material (intervertebral disc). These discs act as shock absorbers to protect the spine. A prolapsed or slipped disc occurs when one of these intervertebral discs is displaced from its normal position and is no longer performing its function properly. In other cases, there may be squashing or compression of a disc. This results in a painful condition, sometimes leading to partial paralysis, which may be caused when the back is bent while lifting, as a result of falling awkwardly, getting up out of a low chair or even through over-energetic dancing.

Rheumatism

This is a painful disorder of joints or muscles not directly due to infection or injury. This rather ill-defined group includes rheumatic fever, rheumatoid arthritis, osteoarthritis, gout and 'fibrositis', itself an ill-defined group of disorders in which muscular and/or joint pain are common factors. There is much evidence to support the fact that stress on the spine, muscles, joints and ligaments during manual handling activities in early life results in rheumatic disorders as people get older.

PRINCIPLES OF SAFE MANUAL HANDLING

Injuries associated with manual handling of raw materials, people, animals, goods and other items are the principal form of injury at work. Such injuries include prolapsed intervertebral (slipped) discs, hernias, ligamental strains and various forms of physical injury. It is essential, therefore, that all people required to handle items be aware of the basic principles of safe manual handling.

The following principles should always be considered in handling activities and in the training of people in manual handling techniques.

Feet Position

Place feet hip breadth apart to give a large base. Put one foot forward and to the side of the object to be lifted. This gives better balance.

Correct Grip

Ensure that the grip is by the roots of the fingers and palm of the hand. This keeps the load under control and permits the load to be better distributed over the body.

Arms Close to Body

This reduces muscle fatigue in the arms and shoulders and the effort required by the arms. It ensures that the load, in effect, becomes part of the body and moves with the body.

Flat Back

This does not mean vertical but at an angle of approximately 15°. This prevents pressure on the abdomen and ensures an even pressure on the vertebral discs. The back will take the weight but the legs do the work.

Chin In

It is just as easy to damage the spine at the top as it is at the bottom. To keep the spine straight at the top, elongate the neck and pull the chin in. Do not tuck the chin on to the chest as this bends the neck.

Use of Body Weight

Use the body weight to move the load into the lifting position and to control movement of the load.

When considering moving a load, make sure your route is unobstructed and that there are no tripping hazards.

Ensure there is an area cleared to receive the load. If your route requires you to wear a safety helmet, eye protection or hearing protection, put them on before you lift.

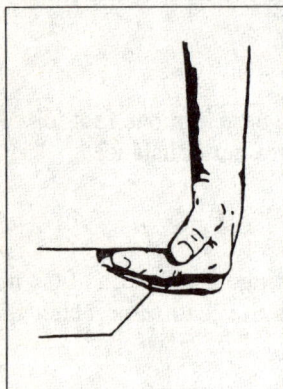

**Figure 19:
Correct Handling
– proper grip**

**Figure 20:
Correct Handling
– straight back**

**Figure 21: Correct Handling
– proper foot positions**

**Figure 22: Correct Handling
– use your bodyweight**

Assessing/Testing the Load

The majority of injuries occur when actually lifting the load. People involved in handling operations should be instructed to assess the following.

1. Are there any rotating or moving parts?
   If so, do not use them to lift with.

2. Is the load too big to handle?
   If so, get help.

3. Is the load too heavy?
   Rock the load. This will give a rough idea of its weight. If it is too heavy, get help.

PERSONAL PROTECTIVE EQUIPMENT (PPE)

The provision and use of the correct PPE is an essential feature of safe manual handling. The following instructions should be incorporated in manual handling training and activities.

Hand Protection

Examine the load for evidence of sharp edges, protruding wires, splinters or anything that could injure the hands. Wear the correct type of glove to prevent hand injury.

Feet Protection

Wear footwear which is suitable for the job:

(a) with steel toe caps to protect the feet against falling objects or if the feet could get trapped under the load;

(b) steel insoles to protect against protruding nails;

(c) soles that will resist heat, oil and acid.

TWO-PERSON LIFT

Use all the principles involved in a one-man lift, with one variation. The leading foot should point in the direction of travel. One person should give the order to lift, ensuring that his partner understands the order. It is vital that there be unison in the movement of both people and the load.

Figure 23: Manual Handling Operations Regulations 1992 – Risk Assessment Checklist (Appendix 1)

Manual handling of loads

EXAMPLE OF AN ASSESSMENT CHECKLIST

Note: This checklist may be copied freely. It will remind you of the main points to think about while you:
- consider the risk of injury from manual handling operations
- identify steps that can remove or reduce the risk
- decide your priorities for action.

SUMMARY OF ASSESSMENT	Overall priority for remedial action: Nil / Low / Med / High*
Operations covered by this assessment:	Remedial action to be taken: ....
	....
Locations:	Date by which action is to be taken: ....
Personnel involved:	Date for reassessment: ....
Date of assessment:	Assessor's name: .... Signature: ....

*circle as appropriate

Section A - Preliminary:

Q1   Do the operations involve a significant risk of injury?   Yes / No*

     If 'Yes' go to Q2. If 'No' the assessment need go no further.

     If in doubt answer 'Yes'. You may find the guidelines in Appendix 1 helpful.

Q2   Can the operations be avoided / mechanised / automated at reasonable cost?   Yes / No*

     If 'No' go to Q3. If 'Yes' proceed and then check that the result is satisfactory.

Q3   Are the operations clearly within the guidelines in Appendix 1?   Yes / No*

     If 'No' go to Section B. If 'Yes' you may go straight to Section C if you wish.

Section C - Overall assessment of risk:

Q   What is your overall assessment of the risk of injury?   Insignificant / Low / Med / High*

     If not 'Insignificant' go to Section D. If 'Insignificant' the assessment need go no further.

Section D - Remedial action:

Q   What remedial steps should be taken, in order of priority?

     i   ....

     ii   ....

     iii   ....

     iv   ....

     v   ....

And finally:

     - complete the SUMMARY above

     - compare it with your other manual handling assessments

     - decide your priorities for action

     - TAKE ACTION.................AND CHECK THAT IT HAS THE DESIRED EFFECT

Figure 23: Manual Handling Operations Regulations 1992 – Risk Assessment Checklist (Appendix 1) (*continued*)

<u>Section B</u> - More detailed assessment, where necessary:

Questions to consider: (If the answer to a question is 'Yes' place a tick against it and then consider the level of risk)	Level of risk: (Tick as appropriate)			Possible remedial action: (Make rough notes in this column in preparation for completing Section D)
	Yes	**Low**	**Med**	**High**

The tasks - do they involve:
- ◆ holding loads away from trunk?
- ◆ twisting?
- ◆ stooping?
- ◆ reaching upwards?
- ◆ large vertical movement?
- ◆ long carrying distances?
- ◆ strenuous pushing or pulling?
- ◆ unpredictable movement of loads?
- ◆ repetitive handling?
- ◆ insufficient rest or recovery?
- ◆ a workrate imposed by a process?

The loads - are they:
- ◆ heavy?
- ◆ bulky/unwieldy?
- ◆ difficult to grasp?
- ◆ unstable/unpredictable?
- ◆ intrinsically harmful (eg sharp/hot?)

The working environment - are there:
- ◆ constraints on posture?
- ◆ poor floors?
- ◆ variations in levels?
- ◆ hot/cold/humid conditions?
- ◆ strong air movements?
- ◆ poor lighting conditions?

Individual capability - does the job:
- ◆ require unusual capability?
- ◆ hazard those with a health problem?
- ◆ hazard those who are pregnant?
- ◆ call for special information/training?

Other factors -
Is movement or posture hindered by clothing or personal protective equipment?

Deciding the level of risk will inevitably call for judgement. The guidelines in Appendix 1 may provide a useful yardstick. When you have completed Section B go to Section C.

SUMMARY – CHAPTER 10

1. Manual handling injuries are the greatest cause of lost time at work.

2. The principal injuries associated with manual handling operations are muscle and ligamental strains, hernias and prolapsed intervertebral discs.

3. The principles of safe manual handling should be incorporated in the induction training of employees and operations monitored regularly to ensure correct lifting procedures are used.

4. Personal protective equipment may be necessary for certain manual handling operations.

5. A manual handling risk assessment should be undertaken where manual-handling operations cannot be avoided.

6. A manual handling risk assessment should take account of the tasks, the loads, the working environment and individual capability, together with other factors, such as the effect of personal protective clothing on manual handling.

Part 4

SAFETY TECHNOLOGY

CHAPTER 11

MACHINERY, EQUIPMENT AND HAND TOOLS

MACHINERY HAZARDS

Many machines, including new machines, incorporate hazards in their basic design. These hazards can be classified as follows.

Traps

Traps can take a number of forms:

(a) reciprocating trap: these may have an up and down motion, e.g. presses; at the point where the injury occurs, the limb is stationary;

(b) shearing trap: these have a guillotine effect;

(c) in-running nips: these are to be found on rollers, conveyors and gears.

Figure 24: Traps

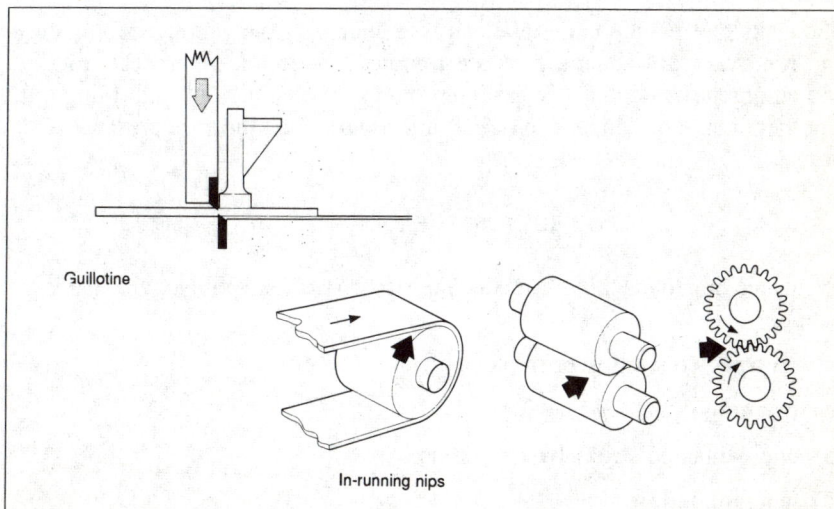

Guillotine

In-running nips

Entanglement

The risk of entanglement of hair, clothing and limbs in, for instance, revolving shafts, line shafts, chucks and drills.

Figure 25: Examples of Risk of Entanglement

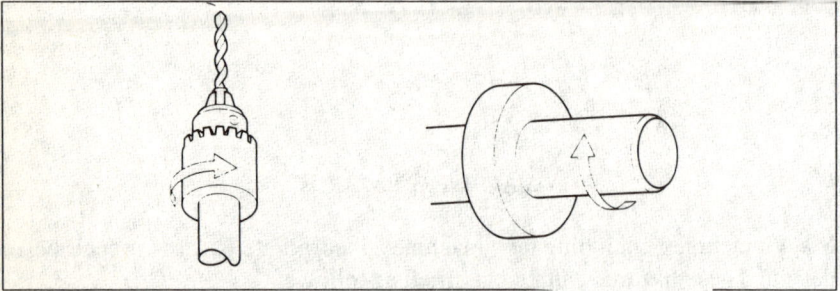

Ejection

The emission or throwing off of particles from a machine, e.g. abrasive wheels, disintegration of swarf on a lathe.

Contact

Contact with a machine at a particular point can cause injury, e.g. heat, temperature extremes, sharp projections, as in plastic moulding machines, circular saws.

GENERAL CIRCUMSTANCES INVOLVING OPERATORS AND OTHERS

The tasks that people undertake can be a source of danger, e.g. job loading and removal, tool changing, waste removal, operation of process, routine and emergency maintenance, gauging, breakdown situations and trying out. Another source of danger can be associated with unauthorised presence and/or use.

SPECIFIC EVENTS LEADING TO INJURY

Events leading to machinery-related injuries vary considerably. Typical events include:

(a) unexpected start-up or movement;

(b) reaching into a feed device;

(c) uncovenanted stroke by a machine;

(d) machine failure.

MACHINERY GUARDS

Fundamentally, there are five specific forms of machinery guard. In many cases they are linked with a safety device.

Fixed Guard

This is a guard which has no moving parts associated with it, or dependent on the mechanism of any machinery, and which, when in position, prevents access to a danger point or area.

Figure 26: Fixed Guard to Transmission Machinery

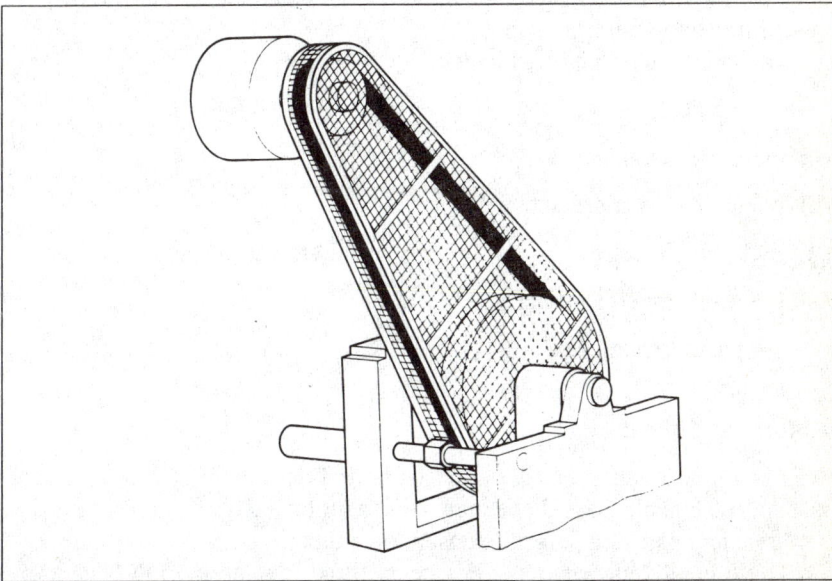

This form of guard is designed to prevent all access to the dangerous parts of the machine and is principally used to cover non-operational parts. Many fixed guards are sold castings, sheet metal (minimum 18 SK – 1.22m), perforated or expanded metal (minimum 17 SWG), 'Weldmesh' (minimum 14 SWG), safety glass panels or polycarbonate panels. Wood as a guard material is not recommended, except where there may be a risk of electric shock.

Interlocking Guard

This is a guard which has a movable part so connected with the machinery ensures that:

(a) the part(s) of the machinery causing danger cannot be set in motion until the guard is closed;

(b) the power is switched off and the motion braked before the guard can be opened sufficiently to allow access to the dangerous parts;

(c) access to the danger point or area is denied while the danger exists.

A guard is also defined as a moving guard which, in the closed position, prevents all access to the dangerous parts, e.g. the control gear for starting up cannot be operated until the guard is fully closed, and the guard cannot be opened until the dangerous parts are at rest.

For a true interlock system, everything must be at rest before the guard or gate can be opened. Some interlocks control only the power supply, and others, the power supply and the movement. In order to achieve the same level of safety as with fixed guards, the reliability and maintenance of interlocking guards are significant.

Methods of interlocking include:

(a) mechanical;

(b) electro-mechanical;

(c) pneumatic (compressed air);

(d) hydraulic (electro) – use of hydraulic fluid to vary pressure;

(e) key exchange (electrical);

(f) simple electrical.

Automatic Guard

This is a guard which is associated with, and dependent upon, the mechanism of the machinery and operates so as to remove physically from the danger area, any part of a person exposed to danger.

These guards incorporate a device so fitted in relation to the dangerous parts that the operator is automatically prevented from contacting them, e.g. heavy power presses, press brakes, paper-cutting guillotines. The guard is independent of the operator.

The function of an automatic guard is to remove the operator from the dangerous parts of the machine by means of a moving barrier or arm. There is some degree of risk in that the operator can be injured by the moving barrier, and this type of guard is only suitable for large slow-moving barriers as on presses. These guards operate on a side-to-side, sweep away or push out motion.

Automatic guards do have a number of disadvantages:

(a) risk of injury to the operator as a result of the sweep-away motion;

(b) the linkages to the motion must be rigidly connected as they can become loose through constant use;

(c) when the linkages become worn the guard is often racing the tools to maintain safe operation (and can lose!);

(d) they need extensive careful maintenance and frequent inspection.

Distance Guard

This is a guard which does not completely enclose a danger point or area but which places it out of normal reach. This may incorporate a tunnel, fixed grill or rail positioned at sufficient distance so that access to the moving parts cannot be gained except by a deliberate unsafe act.

Figure 27: Tunnel guard for a Metal-cutting Machine

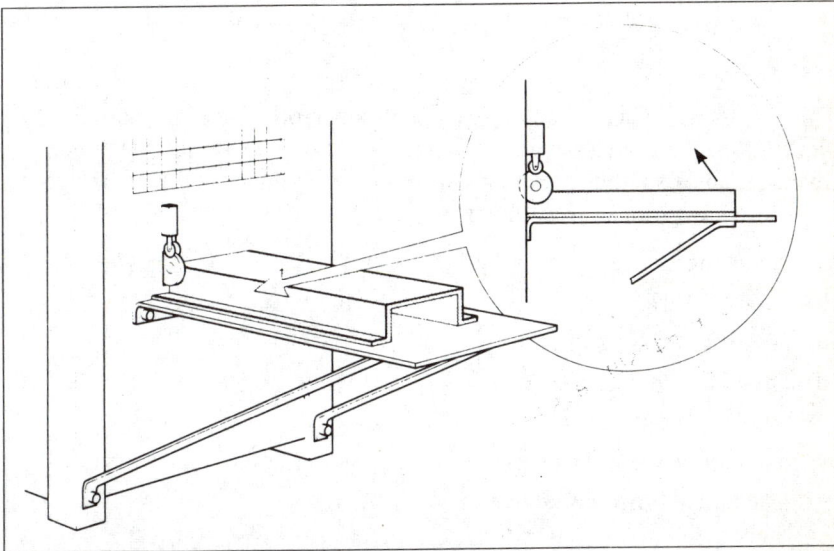

Adjustable Guard

This is a guard incorporating an adjustable element which, once adjusted, remains in that position during a particular operation. This is the least reliable form of guard in that it requires the operator to make adjustments to it prior to operating the machine. Adjustable guards feature particularly in woodworking machinery, e.g. circular saws and band saws.

SAFETY DEVICES

A safety device is a "protective appliance, other than a guard, which eliminates or reduces danger before access to a danger point or area can be achieved". Most safety devices operate on a trip system.

Figure 28: Adjustable Guard to a Band-saw Blade

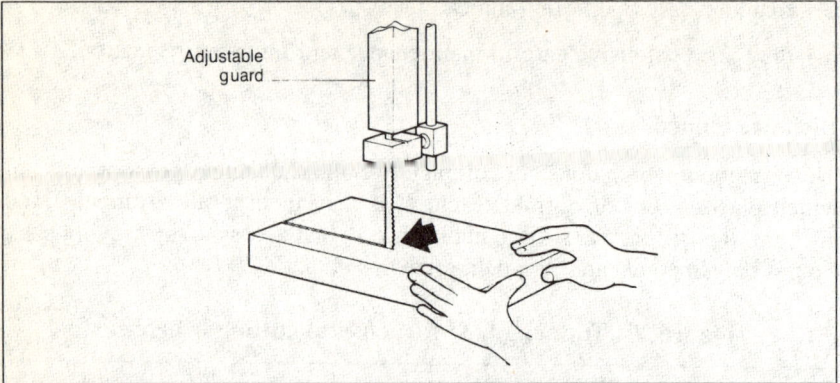

Trip Devices

This is a means whereby any approach by a person beyond the safe limit of working machinery causes the device to actuate and stop the machinery or reverse its motion, thus preventing or minimizing injury at the danger point.

There are various forms of trip device:

(a) mechanical;

(b) photo-electric;

(c) pressure sensitive mat:

(d) ultrasonic device;

(e) two-hand control device;

(f) over-run device;

(g) mechanical restraint device.

Figure 29: Safety Trip bar on a Horizontal Two-roll Mill

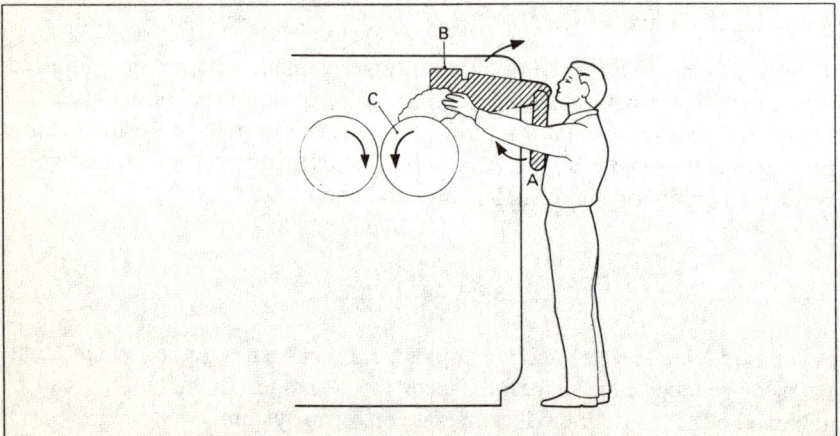

Figure 30: Two-hand Control on a Clicking Press

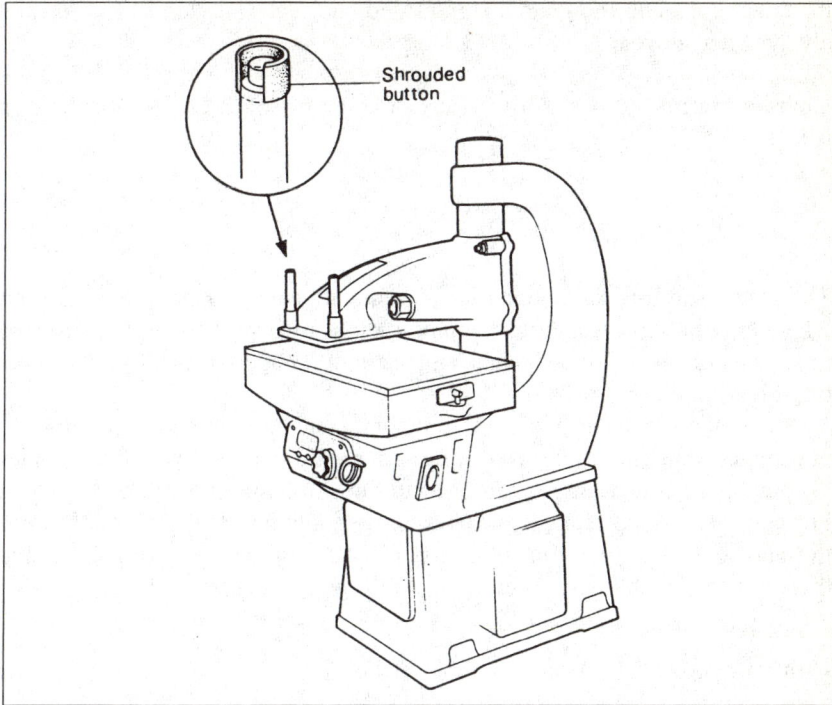

Shrouded
button

Safety Mechanisms

The detailed design of the mechanism controls the safety of the operator. Any consideration of safety mechanisms should include the following objectives and requirements.

Reliability Given the conditions a component is subjected to over a period of time, it must perform in a reliable way. Warning systems must also be reliable to the extent that the operate for the purposes for which they were designed and should be reliable when exposed to oil, vibration, shock, water, etc.

Precise operation The mechanism should operate positively, e.g. precise linkage between rams and guards. The transmission angle on linkages must be minimal and control over wear on linkages is essential.

Protection against operator abuse and misuse Abuse is associated with the operator trying to open the guard before it is due to open, causing wear, and as a result of harsh treatment. Misuse, on the other hand, is a calculated attempt to defeat the safety mechanism. Mechanisms must, therefore, be designed to prevent both abuse and misuse.

Fail-safe   When the component fails, it must do so in such a way that the machine stops and the guards stay closed, and not vice versa. This cannot always be achieved.

Correct method of assembly   Correct assembly of the safety mechanism is vital.

THE SAFETY OF HAND TOOLS

The abuse and misuse of hand tools frequently result in injuries, many of which are of a serious nature, e.g. amputations of fingers, blinding, severing of arteries as a result of deep cuts, and account for approximately 10 per cent of all lost-time injuries.

As with other forms of work equipment, hand tools should be maintained in efficient state, in efficient working order and in good repair. This implies the need for frequent inspection of hand tools to ensure their relative safety. Furthermore, the correct use of hand tools should be ensured through training and regular supervision of users. Hand tools used by contractors and their employees should also be subject to regular examination.

Hand Tool Inspections

A number of points should be considered when examining hand tools.

Chisels

'Mushroomed' chisel heads are a frequent cause of blinding and eye injuries and any mushrooming should be removed through grinding. Chisel heads should be kept free from dirt, oil and grease.

Hammers

The shaft should be in sound condition and soundly fixed to the head. Where the shaft is split, loose to the head, broken or damaged, it should be replaced. Chipped, rounded or badly worn hammer heads should not be used, and heads should be kept free of oil and grease.

Files

A file should never be used without a handle, and the handle should be in sound condition. Evidence of chips and other signs of damage indicate a file could be dangerous.

Figure 31: Is the Machine Safe for Use at Work?

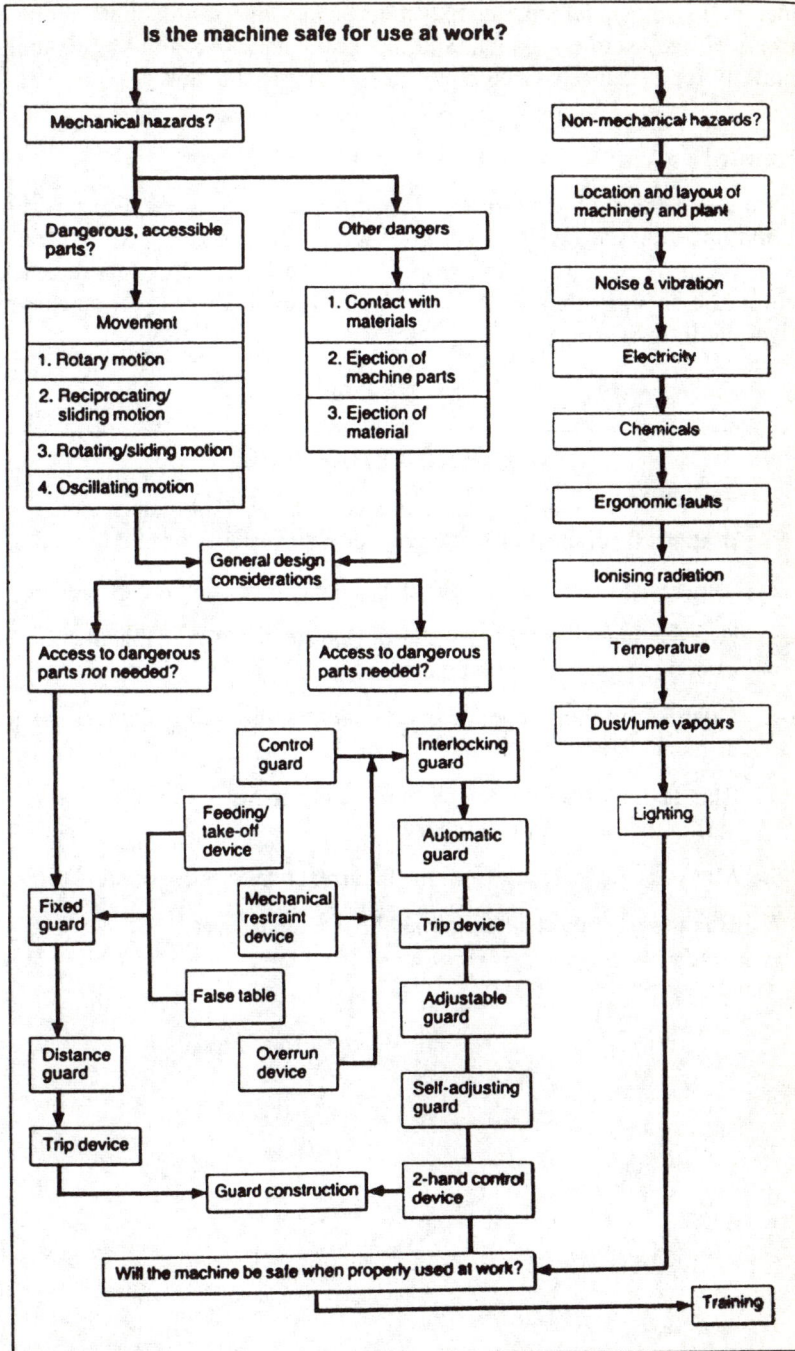

Spanners

Open-end spanners which are splayed or box spanners with splits should be discarded. Adjustable spanners and monkey wrenches should be examined regularly for evidence of free play and splaying of the jaws.

Screwdrivers

Handles and tips should be in sound condition and worn-ended screwdrivers should never be used. A screwdriver should never be used as a chisel and when using a screwdriver, the work should be clamped or secured, never held in the hand. Employees must be trained to use the correct size screwdriver at all times.

SUMMARY – CHAPTER 11

1. The principal machinery hazards are traps, entanglement, ejection of items from machines and contact with surfaces or moving parts.

2. The actual tasks that people undertake can be a source of danger.

3. A wide range of machinery guarding systems are available; the most effective guard is a fixed guard.

4. Guards commonly operate in conjunction with safety devices, such as trip devices.

5. The design of safety devices and mechanisms is significant in ensuring the safety of machinery.

6. Many accidents result from the incorrect use or abuse of hand tools.

7. Hand tools should he examined on a regular basis.

CHAPTER 12

FIRE PREVENTION AND PROTECTION

Every year fire and its effects cause substantial losses of life and property. It is essential that everyone at work is familiar with fire procedures and the measures to prevent fire.

WHAT IS 'FIRE'?

'Fire' can be defined in several ways:

(a) a spectacular example of a fast chemical reaction between a combustible substance and oxygen accompanied by the evolution of heat;

(b) a mixture in gaseous form of a combustible substance and oxygen with sufficient energy put into the mixture to start a fire;

(c) an unexpected combustion generating sufficient heat or smoke resulting in damage to plant, equipment, goods and/or buildings.

PRINCIPLES OF COMBUSTION

In order to appreciate the principles of fire prevention, it is necessary to have a broad understanding of the principles of combustion. The three requirements for a fire to start and continue are: the presence of fuel to burn; an ignition source of sufficient energy to set the fuel alight and air or oxygen to maintain combustion. If one of these three components is removed, combustion cannot take place.

Figure 32: The Fire Triangle

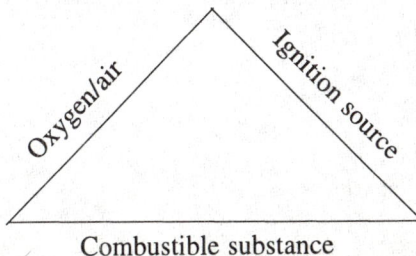

Combustible substance

Elements of Fire

Oxygen/Air

A fire always requires oxygen for it to take place or, having commenced, to continue. The chief source of oxygen is air, which is a mixture of gases comprising nitrogen (78 per cent) and oxygen (21 per cent). The remaining 1 percent is made up of water vapour, carbon dioxide, argon and other gases. A number of substances can be a source of oxygen in a fire, e.g. oxidising agents. These are substances which contain oxygen that is readily available under fire conditions and include sodium chlorate, hydrogen peroxide, nitric acid and organic peroxides. A third source of oxygen is the combustible substance itself, e.g. ammonium nitrate.

Combustible Substance

This is the second requirement for fire and includes a large group of organic substances, i.e. those with carbon in the molecule (e.g. natural gas (methane), butane, petrol, plastics, natural and artificial fibres, wood, paper, coal and living matter). Inorganic substances, i.e. those not containing carbon in the molecule, are also combustible (e.g. hydrogen, sulphur, sodium, phosphorus, magnesium and ammonium nitrate).

Ignition Source

This is the energy that has to be applied to the oxygen/fuel mixture to start the fire. Usually this energy is in the form of heat, but not necessarily. The heat can simply be that contained in the combustible substance. This is often the source of ignition energy when hot fuel leaks from a pipe and fires, but it can be generated by friction, such as striking a match against sandpaper or a hot bearing in a machine. Electrical energy in the lightning of a thunder storm, or when an electrical contact, such as a switch, is made or broken, would also qualify.

THE MAIN CAUSES OF FIRE AND FIRE SPREAD

Various studies by the Fire Protection Association of a range of industrial fires have indicated the following as the principal sources of fire in production and storage areas.

Production Areas

1. Heat-producing plant and equipment.
2. Frictional heat and sparks.

3. Refrigeration plant.
4. Electrical equipment.

So setting fire to:
    (a) materials being processed;
    (b) dust;
    (c) waste and packing materials.

Storage Areas

1. Intruders, including children.
2. Cigarettes and matches.
3. Refuse burning.
4. Electrical equipment.

So setting fire to:
    (a) stored goods;
    (b) packing materials.

Ignition Temperature

Generally, fire is spread by a range or combination of factors, namely through heat transmission, direct contact and/or through the release of flammable gases or vapours. In all cases, some form of ignition source must be present which is sufficient to create the energy necessary to raise a volume of combustible material to its ignition temperature. This is the temperature at which a small amount of combustible material (fuel) will spontaneously ignite in a given atmosphere and continue to burn without any further input of heat. There must be specific conditions present for ignition to take place, as seen with the 'Fire Triangle', (see Figure 32 on p. 141) namely the appropriate temperature, the right source of ignition and an appropriate mixture of combustible material and oxygen.

Sources of Ignition

Classic sources of ignition in both domestic and industrial premises include:

(a) **electrical equipment:** arcing, which results in the production of sparks, and hot surfaces produced by defective electrical equipment, are a common source of ignition;

(b) **spontaneous ignition:** when some liquids are heated or sprayed on to a very hot surface, they may ignite spontaneously without an ignition source actually present;

(c) **spontaneous combustion:** when materials react with oxygen an *exothermic reaction* takes place, i.e. one emitting heat, and with materials which readily oxidise, there may be some degree of heat accumulation which eventually causes the material to ignite or burst into flames;

(d) **smoking by employees and others:** in many work situations, can be a source of fire, principally from discarded cigarette ends and matches, but also from smoking in areas where flammable materials are stored or where flammable vapours may be present;

(e) **friction:** sparks, sufficient to act as a source of ignition, can be created by friction between surfaces, e.g. where the moving part of a machine comes into contact with a fixed part, or two moving surfaces may rub or slide together during routine machine operation;

(f) **hot work:** such as welding, soldering, hot cutting and brazing, can be a source of ignition, particularly where flammable vapours may be present; the operation of a 'permit to work' system may be necessary in high risk situations;

(g) **static electricity:** in situations where electrostatic charging is produced by induction or friction, the charge, in the form of static electricity, can be carried away from the point of origin and, in the event of accumulation of charge, sparks can be produced;

(h) **vehicle maintenance and parking areas:** diesel and petrol-operated engines, vehicle emissions and hot surfaces of, for instance, exhaust systems, can be a source of ignition;

(i) **open flame sources:** can be encountered in workplaces, e.g. boilers, furnaces, portable heating appliances and pilot lights;

(j) **lightning:** in limited cases can be a source of ignition and this may require the installation of lightning protection by direct earthing.

THE SECONDARY EFFECTS OF FIRE

Whilst the primary effects of fire, in most cases, can cause loss of life and substantial damage, the secondary effects must also be considered in any fire protection strategy. These secondary effects can include:

(a) 'smoke-logging' of buildings, which may make them uninhabitable for a period of time;

(b) some reduction in structural stability of adjacent buildings;

(c) damage to services, including engineering, electrical systems, pressure systems, pipework and ancillary systems;

(d) deterioration of wall and ceiling surfaces which may require subsequent redecoration;

(e) deterioration in the condition of stored final products which may render such products unsaleable;

(f) dust explosions in certain types of industrial plant, such as spray drying plant, milling and grinding plant.

THE EFFECTS OF FIRE ON STRUCTURAL ELEMENTS

The structural elements of a typical building include stone, bricks, concrete, timber, glass, steel, plastics, plaster, plasterboard, asbestos-cement products and a range of surface coatings. The effects of fire on all or some of these materials, combined together to form a structure, is complicated and, in some cases, unpredictable. The majority of these elements, other than timber and plastics, are of an incombustible nature. However, all substances respond to heat with different results.

Stone, Brick and Concrete Items

Stone and brick tend to be unaffected by fire other than flaking of the surface. Concrete, on the other hand, suffers dehydration and tends to shrink at around 400°C, with breaking down of the cement and total disintegration of the concrete at around 800°C.

Glass Items

Glass expands on heating resulting in initial cracking and subsequent bursting out of items, such as windows and glass panelled doors. At around 800°C, glass will actually melt.

Plastic Items

Plastic items including pipework of various types, coatings to windows and doors, and various forms of internal finish. Much will depend upon the type of plastic as to the effect, but all plastics will burn. Burning droplets of plastic falling from, for instance, a roof of a building to another can rapidly spread fire to other areas below.

Steel Structural Items

Steel is an excellent conductor of heat. It rapidly expands and will become ductile at between 550-650°C. The expansion can result in distortion of other structural items, such as concrete floors and walls, thereby increasing the rate at which a structure becomes unstable.

Timber Items

All timber burns, the rate of burning depending on the type of timber, i.e. hardwood or softwood, the intensity of heat applied and whether on a continuous basis. The ash formed on the surface acts as an insulating barrier and can slow down the rate of burning.

Surface Coatings

Coatings to internal walls, ceilings and floors vary considerably in their combustibility. Several layers of gloss paint on a wall, for instance, burn well once the right temperature has been reached. This causes softening of the paint surface and allows the fire to spread more rapidly across the painted surface.

Roof Coverings

Typical roof coverings are slates, tiles, asbestos-cement sheetings and bitumen felt. Slates and tiles tend to withstand high levels of temperature. Asbestos-cement sheeting becomes friable and disintegrates noisily in a fire as a result of the rapid decomposition of the cement binder. Bitumen-impregnates felt burns rapidly and, as with plastics, burning droplets can rapidly spread the fire to other parts of the building.

PRINCIPLES OF FIRE SPREAD AND CONTROL

Any fire will continue to spread under the right conditions. These are:

(a) the presence of combustible fuel to burn;

(b) sufficient air with an appropriate oxygen content;

(c) a continuing source of ignition from the existing fire.

Fire spread control, therefore, is directed at eliminating one or more of the above elements which maintain the fire. The ultimate objective is the extinction of the fire, using one or more of the methods listed below.

Starvation

There are three ways that starvation can be achieved:

(a) take the fuel away from the fire;

(b) take the fire away from the fuel;

(c) reduce the quantity or bulk of the fuel.

The first is achieved every day on a gas hob when the tap is turned off. For industrial installations this means isolating the fuel feed at the remote isolation valve. Examples of taking the fire away from the fuel include breaking down stacks and dragging away the burning debris. Breaking down a fire into smaller units is an example of reducing the quantity or bulk of the fuel.

Smothering

Smothering can be achieved by:

(a) allowing the fire to consume the oxygen while preventing the inward flow of more oxygen;

(b) adding an inert gas to the burning mixture.

Wrapping a person, whose clothing is on fire, in a blanket is an example of smothering. Other examples include pouring foam on top of a burning pool of oil or putting sand on a small fire. A danger inherent in extinguishing fires by smothering occurs when the fire is out but everything is still hot. Any inrush of oxygen, caused by disturbing the foam layer or by opening the door to a room, could result in re-ignition as there may still be sufficient energy in the form of sensible heat present.

If an inert gas is to be used as an extinguisher, carbon dioxide or halogenated hydrocarbons, such as BCF, are suitable. Alternatively, nitrogen can be used, and is in fact more common for petrochemical plant fires. If a flammable gas pipeline leaks and the escaping gas ignites, nitrogen blanketing can be achieved by injecting nitrogen into the gas downstream of the release. Smothering is only effective when the source of oxygen is air. It is totally ineffective when the burning substance contains oxygen, such as ammonium nitrate.

Cooling

This is the most common means of fighting a fire, water being the most effective and cheapest medium. For a fire to be sustained, some of the heat output from the combustion is returned to the fuel, providing a continuous source of ignition energy. When water is added to a fire, the heat output serves to heat and vaporize the water, that is, the water provides an alternative heat sink. Ultimately, insufficient heat is added to the fuel and continuous ignition ceases. In order to assist rapid absorption of heat, water is applied to the fire as a spray rather than a jet, the spray droplets being more efficient in absorbing heat than the stream of water in a jet.

Another example of heat absorption is provided by dry chemical extinguishers. These are very fine powders which rapidly absorb heat. Dry powders can also be used for the smothering method.

FIRE ALARM SYSTEMS

A method of giving warning of fire is required in commercial, industrial and public buildings. The purpose of a fire alarm is to give an early warning of a fire in a building:

(a) to increase the safety of occupants by encouraging them to escape to a place of safety;

(b) to increase the possibility of early extinction of the fire thus reducing the loss of or damage to the property.

BS 5839: Part 1:1988 lays down guidelines to be followed for the installation of fire alarm systems. In larger buildings this may take the form of a mains operated system with breakglass alarm call points, an automatic control unit and electrically-operated bells or sirens. In small buildings it would be reasonable to accept a manually operated, dry battery or compressed air-operated gong, klaxon or bell. To avoid the alarm point being close to the seat of a fire, duplicate facilities are necessary.

FIRE INSTRUCTIONS

A fire instruction is a notice informing people of the action they should take on either:

(a) hearing the alarm;

(b) discovering a fire.

OTHER REQUIREMENTS

In addition to displaying fire instructions, people:

(a) should receive training in evacuation procedures, i.e. fire drills, at least once quarterly;

(b) the alarm should be sounded weekly.

It is advantageous to have key personnel trained in the correct use of fire appliances.

SUMMARY – CHAPTER 12

1. Fire is a principal causative factor in loss of life and property every year.

2. For combustion to take place three elements must be present: air, a fuel to burn and an ignition source. If any one of these elements is absent combustion cannot take place.

3. The main causes of fire and fire spread in a place of work must be considered.

4. The secondary effects of fire can, in many cases, be more significant in terms of loss.

5. The effects of fire on structural elements must be considered in the design of buildings.

6. Any fire protection strategy should take into account the principles of fire spread control.

7. Starvation, smothering and cooling are the three principal methods of fire spread control.

8. Fire alarm systems are required in most places of work.

9. Fire instructions must be displayed in places of work.

10. Employees should receive training in the correct use of fire appliances and action to be taken in the event of the alarm being sounded.

CHAPTER 13

ELECTRICAL SAFETY

THE PRINCIPLE HAZARDS

Hazards associated with the use of electricity can broadly be divided into two categories, namely the risk of injury to people and the risk of fire and/or explosion.

INJURIES TO PEOPLE

Human injury is associated with shock, burns, physical injuries from explosions, microwaves, accumulators and batteries and eye injuries.

Electric Shock

This is the effect produced on the body and, in particular, the nervous system, by an electric current passing through it. The effect varies according to the strength of the current which, in turn, varies with the voltage and the electrical resistance of the body. The resistance of the body varies according to the points of entry and exit of the current and other factors, such as body weight and/or the presence of moisture.

Ohm's Law

$$R \text{ (Resistance in ohms)} \quad = \quad \frac{E \text{ (Pressure in volts)}}{I \text{ (Current in amps)}}$$

Table 3: Typical Responses to Current/Voltage

Voltage	Response	Current
15 volts	Threshold of feeling	0.002-0.005 amps
20-25 volts	Threshold of pain	—
30 volts	Muscular spasm (non-release)	0.015 amps
70 volts	Minimum for death	0.1 amps
120 volts	Maximum for 'safety'	0.002 amps
200-240 volts	Most serious/fatal/accidents	0.2 amps

A common cause of death is ventricular fibrillation (spasm) of the heart muscle which occurs at 0.05 amps. The vascular system ceases to function and the victim dies of suffocation.

Remember – IT'S THE CURRENT THAT KILLS

$$\text{Current (amps)} = \frac{\text{Voltage}}{\text{Resistance}}$$

FIRST AID

First aid for a victim of electric shock must be cardiac massage plus mouth-to-mouth resuscitation until normal breathing and the heart action return. A victim who is 'locked on' to a live appliance must not be approached until the appliance is electrically dead.

Burns

A current passing through a conductor produces heat. Burns can be caused by contact with hot conductors or by the passage of a current through the body at the point of entry and exit. Electric arcing from short circuits may also cause burns.

Explosions

Electrical short circuit or sparking from the electrical contacts in switches or other equipment is a common cause of explosions and subsequent human injury or death. This presupposes the presence of a flammable atmosphere, e.g. vapour, dust or gas.

Eye Injuries

These can arise from exposure to ultraviolet rays from accidental arcing in a process such as welding.

Microwave Apparatus

Microwaves can damage the soft tissues of the body.

Accumulators and Batteries

Hydrogen gas may be produced as a by-product of battery charging which can cause explosive atmospheres with the risk of burns.

Electricity is a common source of ignition for major fires. Some insulating materials and materials used for electrical connections may be flammable and can give rise to small fires in switchgear, distribution boxes or electricity sub-stations. The risk of losses from fire increases when these local fires go undetected and result in major fires.

Sources of electrical ignition include:

(a) **sparks:** between conductors or conductor and earth;

(b) **arcs:** are a larger and brighter discharge of electrical energy and are more likely to cause a fire;

(c) **short circuits:** arise when a current finds a path from live to return other than through apparatus, resulting in high current flow, heating of conductors to white heat and arcing;

(d) **overloading:** where too much current flows causing heating of conductors;

(e) **old and defective/damaged wiring:** through breakdown of the insulation resulting in a short circuit, or the progressive use of more equipment on an old circuit resulting in overloading.

PRINCIPLES OF ELECTRICAL SAFETY

The prime objective of electrical safety is to protect people from electric shock, and also from fire and burns, arising from contact with electricity. There are two basic preventive measures against electric shock:

(a) **protection against direct contact:** e.g. by providing proper insulation for parts of equipment liable to be charged with electricity;

(b) **protection against indirect contact:** e.g. by providing effective earthing for metallic enclosures which are liable to be charged with electricity if the basic insulation fails for any reason.

When it is not possible to provide adequate insulation as protection against direct contact, a range of measures is available, including protection by the use of barriers or enclosures, and protection by position, i.e. placing live parts out of reach.

Earthing

This implies connection to the general mass of earth in such a manner as will ensure at all times an immediate discharge of electrical energy without danger. Earthing, to give protection against indirect contact with electricity, can be achieved in a number of ways, including the connection of extraneous conductive parts of premises (radiators, taps, water pipes) to the main earthing terminal of the electrical installation. This creates an equipotential zone and eliminates the risk of shock that could occur if a person touched two different parts of the metalwork liable to be charged, under earth fault conditions, at different voltages.

When an earth fault exists, such as when a live part touches an enclosed conductive part, e.g. metalwork, it is vital to ensure that the electrical supply is automatically disconnected. This protection is brought about by the use of overcurrent devices, i.e. correctly rated fuses or circuit breakers, or by correctly rated and placed residual current The maintenance of earth continuity is vital.

Fuses

A fuse is basically a strip of metal of such size as would melt at a predetermined value of current flow. It is placed in the electrical circuit and, on melting, cuts off the current to the circuit. Fuses should also be of a type and rating appropriate to the circuit and the appliance it protects.

Circuit Breakers

These devices incorporate a mechanism which trips a switch from the *ON* to *OFF* position if an excess current flows in the circuit. A circuit breaker should be of the type and rating for circuit and appliance it protects.

Earth Leakage Circuit Breakers (residual current devices)

Fuses and circuit breakers do not necessarily provide total protection against electric shock. Earth leakage circuit breakers provide protection against earth leakage faults, particularly at those locations where effective earthing cannot, necessarily, be achieved.

Reduced Voltage

Reduced voltage systems are another form of protection against electric shock, the most commonly used being the 110 volt centre point earthed system. In this system the secondary winding of the transformer providing the 110 volt supply is centre tapped to earth, thereby ensuring that at no part of the 110 volt circuit can the voltage to earth exceed 55 volts.

Safe Systems of Work

When work is to be undertaken on electrical apparatus or a part of a circuit, a formally operated safe system of work should always be used. This normally takes the form of a 'permit to work' system which ensures the following procedures:

(a) switching out and locking off the electricity supply, i.e. isolation;

(b) checking by the use of an appropriate voltage detection instrument that the circuit, or part of same to be worked on, is dead before work commences;

(c) high levels of supervision and control to ensure the work is undertaken correctly;

(d) physical precautions, such as the erection of barriers to restrict access to the area, are implemented;

(e) formal cancellation of the permit to work once the work is completed satisfactorily and return to service of the plant or system in question.

SUMMARY – CHAPTER 13

1. The principal hazards associated with electricity are the risk of electrocution and fire.

2. Electrical faults are a common source of ignition for major fires.

3. The principles of electrical safety are insulation, earthing and isolation.

4. A range of devices are available to ensure electrical protection.

5. Work on electrical equipment will, in most cases, require the operation of an established safe system of work, such as a permit to work system.

CHAPTER 14

CONSTRUCTION ACTIVITIES

Traditionally, the construction industry has always ranked as one of the more dangerous industries. This has been due to a number of factors, including the nature of the work, such as work at heights and below ground level, weather conditions, the temporary nature of the work, a lack of safety supervision in many cases, and the pressing need to complete projects on time which can result in safety considerations not been given a high priority.

THE PRINCIPAL HAZARDS

The hazards associated with construction activities are extensive. The principal risks and the causative factors of same are outlined below.

Ladders

Falls from ladders; ladders slipping outwards at the base or falling away at the top (the '1out:4up' rule should always be used); use of defective ladders; over-reaching situations.

Falls from Working Platforms

Unfenced and inadequately fenced working platforms; inadequate and defective boarding to working platforms; absence of toe boards.

Falls of Materials

Small objects, such as bricks and hand tools, dropped from a height; poor standards of housekeeping on working platforms; inadequate or absent toe boards and barriers; incorrect assembly of gin wheels for raising and lowering materials; incorrect or careless hooking and slinging of loads; failure to install catching platforms (fans) for falling debris; demolition materials being thrown to the ground.

Falls from Pitched Roofs and through Fragile Roofs

Unsafe working practices; use of inappropriate footwear; failure to provide eaves protection and verge protection; failure to use crawl boards; stacking of materials on fragile roofs.

Falls through Openings in Flat Roofs and Floors

Failure to cover openings or provide edge protection; failure to replace covers/edge protection; covers not clearly marked to indicate floor openings below.

Collapse of Excavations

Failure to support trench excavations; inadequate timbering and shoring; shifting sand situations; presence of water in large quantities, e.g. flash floods; timbering collapses due to materials stacked and equipment located too close to the edge of an excavation; failure to re-instate supports after damage.

Transport

Falls from vehicles not designed to carry passengers, e.g. dumper trucks; crushing by reversing lorries and trucks; poor maintenance of site vehicles, e.g. braking and reversing systems; operation of vehicles and machinery, particularly lifting appliances, such as cranes, hoists and winches, by inexperienced and incompetent persons; overloading of passenger-carrying vehicles; poor standards of driving on site roads; mud on roads; poor housekeeping on roads causing skidding and obstruction to vehicles.

Machinery and Powered Hand Tools

Failure to adequately guard all moving and dangerous parts of machinery, e.g. power take-offs, cooling fans and belt drives; dangerous woodworking machinery, particularly circular saws; portable hand tools with rotating heads, e.g. angle grinders; defective and uninsulated electric hand tools.

Housekeeping

Poor housekeeping levels; trips and falls over debris accumulated during construction.

Fire

Inadequate fire protection measures, often associated with poor site supervision; uncontrolled welding and burning activities; burning of site refuse.

Personal Protective Equipment

Failure to provide and enforce the wearing and use of personal protective equipment, such as safety helmets, full-face protection, eye protection, safety boots, gloves, overalls, etc.

Work over Water and Transport over Water

Failure to provide barriers, life jackets or buoyancy aids equipment; defective and inadequate boats for conveying workplaces; overcrowding of boats.

Work involving Hazardous Substances

Failure to prevent or adequately control risks from exposure to and/or use of hazardous substances, e.g. lead, asbestos; poor levels of personal hygiene; failure to undertake health risk assessments, air monitoring and health surveillance; inadequate prevention or control strategies; inappropriate personal protective equipment, e.g. respiratory protective equipment.

Manual Handling Operations

Failure to provide mechanical handling aids; failure to undertake risk assessments; inadequate information, instruction and training,

Underground Services

Damage to underground services during excavation work; failure to consult existing service plans and establish location of service lines; failure to use cable and service locators; unsafe digging and excavating practices;

Confined Spaces

Risks of asphyxiation and anoxia; failure to operate a safe system of work, e.g. use of permit to work system; inadequate ventilation; failure to provide and use breathing apparatus; inadequate air monitoring; inadequate communication systems.

CONSTRUCTION ACTIVITIES: MAIN PRECAUTIONS

Work above Ground

This entails the use of scaffolds, mobile access equipment and ladders. The following factors should be considered in order to ensure safe working practices.

Basic Scaffolding Requirements

1.   Safe means of access to and egress from working platforms.

2.   All workplaces above ground to be kept safe.

3.   Scaffolds provided at working heights above 2m.

4.   Toe boards, hand rails and intermediate rails fitted and maintained.

5.   Working areas adequately lit.

6.   No materials to be thrown or tipped from working platforms.

7.   Constructed using approved materials in sound condition.

8.   Rigidly constructed as to prevent accidental displacement.

9.   Standards vertical or leaning towards structure; securely fixed and braced.

10. Ledgers and transoms horizontal and securely fixed.

11. Putlogs straight, provided with flat ends and securely fixed.

12. Gangways adequate (minimum 0.44m wide).

13. Working platforms adequate (minimum 0.64m for general work).

14. Stairs fitted with handrails, toe boards on landings.

15. Warning notices displayed and access blocked to partly-erected scaffolds.

16. Careful lowering of items during dismantling.

17. Rigidly connected to building unless constructed as an independent scaffold.

18. Securely supported or suspended or strutted or braced.

Movable Access Equipment

This type of equipment commonly comprises a movable tower formed from scaffold tubes or pre-formed frames. In each case, the tower incorporates a working platform, access by means of an externally fixed ladder or internally-placed raking ladders, and caster wheels at the base, which permit the tower to be moved with ease.

   This equipment is commonly used for high level maintenance work, painting and small-scale building work. The following requirements are necessary to ensure safe working:

1.   Working platform secure, completely boarded and fitted with toe boards, intermediate rails and hand rails.

2. Height must not exceed three times the smaller base dimension.

3. Outriggers may be necessary to increase stability during windy conditions.

4. Diagonal bracing on all four elevations and horizontally.

5. Casters at four corners securely fixed and fitted with a brake.

6. Moved with great care by pushing/pulling at base level.

7. No person, equipment or materials to be on platform during movement.

See further BS5973:1981 code of practice for access and working scaffolds and special scaffold structures in steel.

Ladders

Some safety precautions cover the actual construction and use of ladders are:

1. Sound construction and maintenance.

2. No use of defective ladders.

3. Wooden stiles and rungs to have grain running lengthwise.

4. Ladders not to be painted or treated as to hide defects; treatment with clear preservative is acceptable.

5. Wooden ladders fitted with reinforcing ties.

6. Equally and properly supported on each stile.

7. Securely fixed near its upper resting place or at lower end or footed.

8. Must rise at least 1m above the landing place.

9. Landing places every 9.14m (30ft) of vertical distance; fitted with toe boards and hand rails.

10. Openings in landings as small as possible.

11. Folding ladders to have a level and firm footing.

Work below Ground

The principal risks are collapse of an excavation, flooding and people, materials and vehicles falling into excavations. Factors for consideration in the support of an excavation are:

(a) the nature of the subsoil;

(b) projected life of the excavation;

(c) work to be undertaken, including equipment used;

(d) possibility of flooding;

(e) depth of the excavation;

(f) number of operators working in the excavation at any one time.

The principal safety requirements are as follows:

(a) an adequate supply of timber and other materials must be available;

(b) barriers to be installed as close as practicable to the edge and maintained in position;

(c) adequate and suitable materials to be used for shoring;

(d) timbering completed as early as practicable, and people to be protected whilst undertaking this work;

(e) timbering to be of good construction, sound material, free from patent defect and of adequate strength;

(f) struts and braces to be adequately secured;

(g) experienced operators to be employed for erection, alteration and dismantling;

(h) excavations and approaches to be well-lit;

(i) no materials to be placed near the edge of an excavation;

(j) steps to prevent premature collapse where excavations may affect the stability of a building;

(k) means for reaching a place of safety, where there is risk of sudden flooding;

(l) means to prevent over-running of vehicles to be installed;

(m) atmosphere to be well-ventilated;

DEMOLITION OPERATIONS

Demolition is the most hazardous operation undertaken in construction activities. The principal hazards are:

(a) falls of men and materials;

(b) collapse of structures;

(c) overloading of floors with debris;

(d) incorrect or unsafe demolition techniques;

(e) explosions in tanks or other confined spaces;

(f) the presence of live electric cables and gas mains;

(g) the presence of dusty, corrosive and poisonous materials and/or atmospheres;

(h) projecting nails, broken glass and cast iron fragments which can cause minor injuries.

The principal precautions entail:

(a) a pre-demolition survey to identify, for instance, the nature and method of construction, previous use, location of services, presence of dangerous substances, cantilevered structures, etc;

(b) isolation of services;

(c) segregation of the area by barriers, control of access, display of warning notices, etc;

(d) installation of fans or catching platforms where necessary;

(e) ensuring provision and use of correct personal protective equipment by operators – safety boots, hard hats, respiratory protection;

(f) use of temporary props where necessary;

(g) effective control when pulling arrangements, a demolition ball, pusher arm and/or explosives are being used;

(h) protection against falling items;

(i) control over access to dangerous areas;

(j) use of scaffolds where manual demolition is undertaken;

(k) protection against falling – safety harnesses, nets, sheets to be used;

(l) no work over open joisting;

(m) prior removal of glass in windows, doors and partitions;

(n) adequately lighting;

(o) competent person to oversee work and make continuing inspections;

(p) express measures to prevent premature collapse;

(q) trained drivers/operators and banksmen whenever mechanical demolition to be undertaken.

COMPETENT PERSONS IN CONSTRUCTION OPERATIONS

The Construction (Lifting Operations) Regulations 1961 require that competent persons undertake the following duties:

(a) supervision of the erection of cranes;

(b) the installation or adjustment of any safety device;

(c) the inspection and testing of safety devices.

Under the Construction (Design and Management) Regulations 1994 competence must be taken into account by:

(a) a client when appointing a planning supervisor;

(b) any person when arranging for a designer to prepare a design;

(c) any person when arranging for a contractor to carry out or manage construction work.

In the case of the Construction (Health, Safety and Welfare) Regulations 1996 competent persons must be appointed for:

1. supervision of:
    (a) the installation or erection of any scaffold and any substantial addition or alteration to a scaffold;
    (b) the installation or erection of any personal suspension equipment or any means or arresting falls;
    (c) erection or dismantling of any buttress, temporary support or temporary structure used to support a permanent structure;
    (d) demolition or dismantling of any structure, or any part of any structure, being demolition or dismantling that gives rise to a risk of danger to any person;
    (e) installation, alteration or dismantling of any support for an excavation;
    (f) construction, installation, alteration or dismantling of a cofferdam or caisson;
    (g) the safe transport of any person conveyed by water to or from any place of work.

2. inspection of places of work as specified in Schedule 7 to the Regulations.

CONSTRUCTION (HEALTH, SAFETY AND WELFARE) REGULATIONS 1996

Schedule 7

Regulation 29(1)

Places of Work requiring Inspection by a Competent Person

Place of Work	*Time of Inspection*
1. Any working platform or part thereof or any personal suspension equipment provided pursuant to paragraph (3)(b) or (c) of Regulation 6.	1. (i) before being taken into use for the first time; and (ii) after any substantial addition, dismantling or other alteration; and (iii) after any event likely to have affected its strength or stability; and (iv) at regular intervals not exceeding seven days since the last inspection.
2. Any excavation which is supported pursuant to paragraphs (1), (2) or (3) of Regulation 12.	2. (i) before any person carries out work at the start of every shift; and (ii) after any event likely to have affected the strength or stability of the excavation or any part thereof; and (iii) after any accidental fall of rock or earth or other material.
3. Cofferdams and caissons	3. (i) before any person carries out work at the start of every shift; and (ii) after any event likely to have affected the strength or stability of the cofferdam or caisson or any part thereof.

MAINTENANCE WORK

Maintenance work may be of a planned and routine nature. Other forms of maintenance may be required in crisis situations.

The principal hazards associated with maintenance work can be classified as below:

1. **Mechanical:** machinery traps, entanglement, contact, ejection; unexpected start-up of machinery.

2. **Electrical:** electrocution, shock, burns, fire.

3. **Pressure:** unexpected pressure releases, explosion.

4. **Physical:** extremes of temperature, noise, vibration; dust and fumes.

5. **Chemical:** gases, fogs, mists, fumes, etc.

6. **Structural:** obstructions, floor openings.

7. **Access:** work at heights, work in confined spaces.

The following precautions should be considered:

1. Safe systems of work.

2. Permit to work systems.

3. Designation of competent persons.

4. Use of method statements.

5. Operation of company contractors' regulations.

6. Controlled areas.

7. Access control.

8. Information, instruction and training.

9. Supervision arrangements.

10. Signs, marking and labelling.

11. Personal protective equipment.

Figure 33: Construction Activities: Health and Safety Checklist

1. HEALTH AND WELFARE

First aid boxes	Responsible person
Ambulance arrangements	Stretcher
Foul weather shelter	Mess room

Clothing storage/changing facilities
Sanitation – urinals, water closets
Drinking water
Emergency procedure

Food heating facilities
Washing facilities
Facilities for rest

2. ENVIRONMENTAL ASPECTS

Access to site and all parts/safe egress
Perimeter signs
Site lighting and emergency lighting
Dust and fume control
Waste storage and disposal
Ventilation arrangements
Adverse lighting
Secondary lighting

Housekeeping, order and cleanliness
Segregation from non-construction
activities
Effective ventilation
Temperature control – indoor
workplaces

3. FIRE PROTECTION

Access for fire brigade appliance
Siting of huts
Flammable refuse storage
Prohibited areas – notices displayed
Flame producing plant and equipment
Vehicle parking arrangements
Storage of flammable substances

System for summoning fire brigade
Space between huts
Fireproof but construction
Specific fire risks
Heaters and heading in huts
Fire appliances
Fire detectors and alarms

4. STORAGE OF MATERIALS

Siting
Storage huts
Compressed gases
Segregation of compressed gases

Stacking
Separation of flammable materials
Hazardous substances
Explosives

5. PLANT, MACHINERY AND HAND TOOLS

Lifting appliances
Electrical equipment
Welding equipment
Maintenance, examination and testing
Guarding and fencing arrangements

Woodworking machinery
Abrasive wheels
Hand tools
Construction, strength and
suitability

6. ACCESS EQUIPMENT AND WORKING PLACES

Scaffolding
Ladders
Trenches/excavations
Fragile material
Means for arresting falls
Competent persons

Overhead cables
Working platforms
Movable access equipment
Falling objects
Personal suspension equipment
Unstable structures

7. COFFERDAMS AND CAISSONS

Design and construction
Strength and capacity
Competent person

Materials
Maintenance

8. PREVENTION OF DROWNING

Drowning risks
Transport over water
Flooding risks

Rescue equipment
Vessels
Work over water

9. RADIOGRAPHY AND RADIOACTIVE MATERIALS

Competent person
Medical supervision arrangements
Control of sealed sources
Dose records

Classified workers
Personal dose monitoring
Records
Personal dose meters/badges

10. SITE TRANSPORT

Cautionary signs and notices
Site layout
Lift trucks
Dangerous vehicles
Controlled access/egress to site
Segregation of traffic routes
Clear traffic routes
Passenger carrying vehicles
Prevention of over-running

Directional signs
Authorised drivers
Mobile access equipment
Unsafe driving
Separate parking areas
Suitability of traffic routes
Towing procedures
Safe loading of vehicles
Emergency routes and exits

11. PERSONAL PROTECTIVE EQUIPMENT

Safety helmets
Gloves/gauntlets
Foul weather clothing
Personal suspension equipment
Safety footwear

Eye/face protection
Respiratory protection
Hearing protection
Donkey jackets

12. DEMOLITION

Pre-demolition survey
Method statement

Competent persons
Asbestos

13. PERSONNEL

General safety training
First aid training

Competent person training
Health surveillance

14. INSPECTION AND REPORTS

Guard rails etc.

Excavations

Means for arresting falls

Welfare facilities

Reporting arrangements

Working platforms

Personal suspension equipment

Ladders

Cofferdams and caissons

ACTION

1. Immediate action

2. Within 7 days

3. Within 28 days

4. Long-term action

_____ Date _____ (Auditor)

SUMMARY – CHAPTER 14

1. Traditionally, construction has always been one of the more dangerous industries.

2. The principal hazards in construction are associated with falls of operators and materials from a height, collapses of excavations, site transport, the use of machinery and powered hand tools, fire and work in confined spaces.

3. Construction (Design and Management) Regulations 1995 lay down the general principles for safety management in construction projects.

4. Site safety inspections should cover work above ground, the use of movable access equipment, including ladders, work below ground and demolition operations in particular.

5. Maintenance work accidents are common, and need a range of safety precautions, such as the operation of 'permit to work' systems, to ensure safe working.

6. Clients have specific duties under the CDM Regulations.

7. Construction activities should be subject to regular inspections and checks.

CHAPTER 15

PERSONAL PROTECTIVE EQUIPMENT

This term includes equipment worn and used by people at work to protect them from both general and specific risks. 'Personal protective equipment' (PPE) is generally defined as meaning "all equipment (including clothing affording protection against the weather) which is intended to be worn or held by a person at work and which protects him against one or more risks to his health or safety, and any addition or accessory designed to meet that objective".

THE RANGE OF PPE

A wide range of PPE is available for use by people at work. It includes:

Head Protection

Industrial safety helmets, various forms of riding helmets, industrial scalp protectors (bump caps), caps and hair nets.

Eye Protection

Safety spectacles, eye shields, safety goggles and face shields.

Face Protection

Face shields which can be hand-held, fixed to a helmet or strapped to the head.

Respiratory Protection

General purpose dust respirators, positive pressure powered dust respirators, helmet-contained positive pressure respirators, gas respirators, emergency escape respirators, air-line breathing apparatus, self-contained breathing apparatus.

Hearing Protection

Ear plugs, ear defenders, muffs and pads, ear valves, acoustic wool.

Skin Protection

Barrier creams and sprays.

Body Protection

One-piece and two-piece overalls, donkey jackets, rubber and PVC-coated aprons, vapour suits, splash-resistant suits, warehouse coats, body warmers, thermal and weather protection overclothing, oilskin overclothing, high visibility clothing, personal buoyancy equipment, e.g. life jackets.

Hand and Arm Protection

General purpose fibre gloves, PVC fabric gauntlets, leather gloves and sleeves. Wrist protectors, chain mail hand and arm protectors.

Leg and Foot Protection

Safety boots and shoes, wellington boots, clogs, foundry boots, anti-static footwear, together with gaiters and anklets.

LIMITATIONS IN THE USE OF PERSONAL PROTECTIVE EQUIPMENT

The use of any form of PPE should, in the majority of cases, be seen either as:

(a) an interim measure until an appropriate 'safe place' strategy, e.g. machine guarding, can be implemented;

(b) the last resort, when all other protection strategies have failed.

Mere provision of PPE is never the perfect solution to protecting people from hazards due to the need for users to wear or use the equipment all the time they are exposed to such hazards. There are a number of reasons why people do not always do this:

(a) it may create discomfort, restrict movement and be difficult to put on or remove;

(b) it may obscure vision;

(c) it may reduce their perception of hazards;

(d) it may be inappropriate to the risk, for example, where unsuitable respiratory protection is provided;

(e) it requires, in many cases, frequent cleaning, replacement of parts, maintenance or some form of regular attention by the user, which he may see as a chore;

(f) some people perceive the use of PPE as unnecessary, a sign of immaturity or yet another management imposition.

SELECTION OF PERSONAL PROTECTIVE EQUIPMENT

A systematic approach to the selection of PPE is essential to ensure that workers at risk are adequately protected. Generally, PPE must be 'suitable', in terms of preventing or controlling exposure to a risk and for the work being undertaken.

When considering the type and form of equipment to be provided, and its relative suitability, the following factors are relevant:

(a) the needs of the user in terms of comfort, ease of movement, convenience in putting on, use and removal, and individual suitability;

(b) the ergonomic requirements and state of health of the persons who may use the PPE;

(c) the capability of the PPE to fit the wearer correctly, if necessary, after adjustments within the range for which it is designed;

(d) the number of personnel exposed to a particular hazard, for instance, noise, dust or risk of hand injury;

(e) the risk or risks involved, the conditions at the place where the exposure to risk may occur and the relative appropriateness of the PPE in protecting operators against, for example, fume and dust inhalation or molten metal splashes;

(f) its relative effectiveness to prevent or adequately control the risk or risks without increasing overall risk;

(g) the scale of the hazard;

(h) standards representing recognised 'safe limits' for the hazard, e.g. HSE Guidance Notes, British Standards;

(i) specific Regulations currently in force;

(j) specific job requirements or restrictions, e.g. work in confined spaces, roof work;

(k) the presence of environmental stressors which will affect the individual wearing or using the equipment, e.g. extremes of temperature, inadequate ventilation, background noise;

(l) the ease of cleaning, sanitization, maintenance and replacement of equipment and/or its component parts.

SUMMARY – CHAPTER 15

1. Personal protective equipment (PPE) includes a range of equipment worn and used by people at work, such as safety boots, eye protection and hearing protection.

2. The provision of PPE is not a perfect solution to providing protection from hazards due to the reluctance of operators to use it.

3. The limitations of PPE should be appreciated.

4. The Personal Protective Equipment at Work Regulations deal with the legal requirements with regard to the selection, provision and use of PPE.

BIBLIOGRAPHY AND FURTHER READING

Secretary of State for Employment, *Health and Safety at Work etc. Act 1974*, HMSO, London (1974).

Health and Safety Executive, *A Guide to the Health and Safety at Work etc. Act 1974: Guidance to the Act*, HMSO, London (1990).

Health and Safety Executive, *Health and Safety Law: What you should know*, HMSO, London (1989).

Health and Safety Executive, *Our Health and Safety Policy Statement writing your Health and Safety Policy Statement: Guide to preparing a Safety Policy Statement for Small Businesses*, HMSO, London (1992).

Health and Safety Executive, *The Health and Safety System in Great Britain*, HMSO, London (1992).

Health and Safety Commission, *Management of Health and Safety at Work: Approved Code of Practice: Management of Health and Safety at Work Regulations 1992*, HMSO, London (1992).

Health and Safety Executive, *Successful Health and Safety Management*, HMSO, London (1991).

Health and Safety Executive, *Five Steps to Risk Assessment*, HSE Information Centre, Sheffield (1998).

Health and Safety Executive, *The Reporting of Injuries, Diseases and Dangerous Occurrences*, HSE Information Centre, Sheffield (1996).

Health and Safety Executive, *Guide to the Reporting of Injuries, Diseases and Dangerous Occurrences Regulations 1995*, HMSO, London (1996).

Health and Safety Executive, *Safe Systems of Work*, HSE Information Centre, Sheffield (1989).

Health and Safety Executive, *Lighting at Work*, HMSO, London (1987).

Health and Safety Executive, *Workplace Health, Safety and Welfare: Approved Code of Practice: Workplace (Health, Safety and Welfare) Regulations 1992*, HMSO, London (1992).

Health and Safety Executive, *Protecting your Health at Work*, HSE Information Centre, Sheffield (1993).

Health and Safety Executive, *Guidance Note EH 22: Ventilation of the Workplace*, HMSO, London (1988).

Health and Safety Executive, *Guidance Note EH 44: Dust in the Workplace*, HMSO, London (1984).

Health and Safety Executive, *Personal Protective Equipment at Work: Guid-*

ance on the Personal Protective Equipment at Work Regulations 1992, HMSO, London (1992).

Health and Safety Executive, *Guidance Note HS(G) 48: Human Factors in Industrial Safety*, HMSO, London (1989).

Health and Safety Commission, *First Aid at Work: Health and Safety (First Aid) Regulations 1981 and Guidance - Approved Code of Practice*, HMSO, London (1990).

Health and Safety Executive, *Manual Handling: Guidance on the Manual Handling Operations Regulations 1992*, HMSO, London (1992).

Health and Safety Executive, *Getting to Grips with Manual Handling*, HSE Information Centre, Sheffield (1993).

British Standards Institution, *Code of Practice: Safeguarding of Machinery (BS 5304)*, BSI, London.

Health and Safety Executive, *Memorandum of Guidance on the Electricity at Work Regulations 1989*, HMSO, London (1989).

Institution of Electrical Engineers, *IEE Regulations for Electrical Installations (The 'Wiring Regulations')*, IEE, Hitchin, Herts.

Health and Safety Commission, *Managing Construction for Health and Safety: Construction (Design and Management) Regulations 1994: Approved Code of Practice*, HSE Books (1995).

Health and Safety Commission, *Designing for Health and Safety in Construction: A Guide for Designers on the Construction (Design and Management) Regulations 1994*, HSE Books (1995).

Health and Safety Commission, *A Guide to Managing Health and Safety in Construction*, HSE Books (1995).

Health and Safety Executive, *Health and Safety for Small Construction Sites*, HSE Books (1995).

Health and Safety Commission, *Approved Guide to the Classification and Labelling of Substances and Preparations Dangerous for Supply: Chemicals (Hazard Information and Packaging for Supply) Regulations 1994: Guidance on Regulations*, HSE Books (1994).

Health and Safety Commission, *Control of Substances Hazardous to Health: Approved Code of Practice: Control of Carcinogenic Substances: Approved Code of Practice: Control of Biological Agents: Approved Code of Practice*, HMSO, London (1994).

Bilsom International, *In Defence of Hearing*, Bilsom International Ltd, Henley-on-Thames (1992).

Lyons, S, *Management Guide to Modern Industrial Lighting*, Butterworths, Sevenoaks (1984).

Stranks, J, *The Handbook of Health and Safety Practice*, Pitman Publishing, London (5th edn, 1998).

Stranks, J, *Human Factors and Safety*, Pitman Publishing, London (1994).

Stranks, J, *Management Systems for Safety*, Pitman Publishing, London (1994).

Stranks, J, *Occupational Health and Hygiene*, Pitman Publishing, London (1995).

Stranks, J, *Safety Technology*, Pitman Publishing, London (1996).

Stranks, J, *Health and Safety Law*, Pitman Publishing, London (2nd edn, 1996).

Stranks, J, *The Law and Practice of Risk Assessment*, Pitman Publishing, London (1996).

Stranks, J, *One Stop Health and Safety*, ICSA Publishing (1996).

Stranks, J, *People at Work: the Human Factors Approach to Health and Safety*, Technical Communications (Publishing) Ltd, Hitchin, Herts. (1996).

Stranks, J, *A Manager's Guide to Health and Safety at Work*, Kogan Page, London (5th edn, 1997).

Confederation of British Industry, *Developing a Safety Culture*, CBI, London (1991).

Beckingsale, A A, *The Safe Use of Electricity*, RoSPA, Birmingham (1976).

INDEX

Risk prevention/control 54
Roof work hazards 156
Routes of entry 98, 99
Rupture 122

Safe access and egress 50
Safe materials 49
Safe person strategies 50, 51
Safe place strategies 49, 50
Safe plant and equipment 49
Safe premises 49
Safe processes 49
Safe systems of work 33, 50, 154
Safety audits 32, 60, 62-68
Safety committees 25, 26, 37-39
Safety culture 44-46
Safety devices 135-138
Safety inspections 32, 60
Safety management requirements 20
Safety mechanisms 137, 138
Safety monitoring 60-80
Safety representatives 9, 21, 24, 25
Safety Representatives and Safety
    Committees Regulations 1977 25,
    26
Safety rules 44
Safety sampling 61, 68, 69
Safety surveys 60
Safety tours 32, 60
Sanitary facilities 90
Scaffolding requirements 158
Secondary cutaneous sensitisers 99
Sensitising substances 97
Separation 104
Severity rate 79
Shared workplaces 20, 21
Shearing trap 131
Sheriff Court 22
Sickness absence rate 79
Skin protection 169
Slipped disc 122
Smoke 103
Social Security (Industrial
    Injuries)(Prescribed Diseases)
    Regulations 1985 99
Space requirements 90
Spontaneous combustion 143
Spontaneous ignition 143

Statements of Health and Safety
    Policy 9, 12, 13, 41-43
Static electricity 144
Statute law 3
Statutes 6
Statutory duty, breach of 4
Stress 54
Structural safety 90
Sub-acute effect 98
Substance 12
Substitution 33, 104
Supervision 33
Supreme Court of Judicature 23
Systemic effect 98

Target organs and systems 101
Temperature 90, 91
Temporary employment 4
Temporary workers 21
Threshold Limit Values 100
Tissue response 101
Torts 4
Total working system 118
Toxic for reproduction (substances)
    97
Toxicology, principles of 98-101
Toxic substance 97
Toxic substances, handling 102
Traffic routes 90
Training 33, 39-41, 56
Training needs 33
Training plans and programmes 39,
    40
Training sessions 40, 41
Traps 131
Trespass 4
Trip devices 136
Two-person lift 125

Underground services 157

Vapours 103
Vehicle movements 156
Ventilation 55, 90, 91, 923, 104-106
Very toxic substance 96
Vibration 92
Vulnerable groups 51